The Rape
of
Ma Bell

The Rape of Ma Bell

The Criminal Wrecking of the Best Telephone System in the World

by Constantine Raymond Kraus
and Alfred W. Duerig

Lyle Stuart Inc.   Secaucus, New Jersey

Published by Lyle Stuart Inc.
120 Enterprise Ave., Secaucus, N.J. 07094
Published simultaneously in Canada by
Musson Book Company,
A division of General Publishing Co. Limited
Don Mills, Ontario

Address queries regarding rights and permissions
to Lyle Stuart Inc., 120 Enterprise Ave.,
Secaucus, N.J. 07094

Manufactured in the United States of America

Library of Congress Cataloging-in-Publication Data

Kraus, Constantine Raymond
    The Rape of Ma Bell : the criminal wrecking of the best telephone
system in the world / by Constantine Raymond Kraus and Alfred W.
Duerig.
        p.   cm.
    Bibliography: p.
    Includes index.
    ISBN 0-8184-0468-X : $19.95
    1. American Telephone and Telegraph Company--Reorganization.
2. Telephone--United States--History. I. Duerig, Alfred W.
II. Title.
HE8846.A55K73 1988
384.6'065'73--dc 19   88-20036
                                                                 CIP

This book is dedicated to the hundreds of thousands of employees of the former Bell System—Ma Bell—who, through their devoted service, made the Bell System a model of technical competence, corporate integrity, and managerial excellence.

Acknowledgments

The authors wish to acknowledge:

The loyal support of their wives, Doris Virginia Kraus and Jean Kohler Duerig. They sustained the authors during their many years of conflict with the changing world of telecommunications during the writing of this book.

The suggestions and comments of Dr. Almarin Phillips, John C. Hower Professor of Public Policy at the Wharton School of the University of Pennsylvania.

The encouragement provided by Mr. Perry Swisher, President of the Idaho Public Utility Commission, and Mr. William Montgomery, former Chairman of the Indiana Public Utility Commission.

The stimulus of Federal Judge Charles R. Richey's decision in the case of Southern Pacific Communications Corporation versus AT&T.

The comments of Myron C. Butler, prolific inventor and specialist in telecommunications.

Dr. William H. Doherty, former Director of Electronic and Television Research in the Bell Laboratories, for his suggestions on the chapter "Ma Bell's House of Wonders."

Raymond G. Ruwell, former member of the Technical Staff in the Bell Laboratories, for his contributions.

Harry B. McCurdy, formerly of the Engineering Department of The Bell Telephone Company of Pennsylvania, for his assistance in reading the manuscript.

The authors pay tribute to John deButts, former Board Chairman of AT&T.

Henry H. Abbott, formerly Department Head in the Bell Laboratories, for his encouragement over many years.

The many retired employees of the Bell System who have expressed their sympathy and encouragement.

The extensive emendations and editing work done by Sandra Stuart, which made the final manuscript more readable for the general public.

Contents

Introduction

"The breakup of America's telephone system, acknowledged to be the most efficient in the world, will affect nearly every aspect of our society. How all this came to pass is a frightening example—at a time when America's industries are in a fight for survival against foreign competitors—of what can happen here to a company recognized as one of our major national, and national defense, assets. It is a company that grew by its own efforts and with its own resources to become the world's largest business, whose Bell Laboratories led us into the information age."

—WALTER H. ANNENBERG, 1983

It was a Tuesday evening in November. Hundreds of thousands of people were pouring from offices and businesses on their nightly trek home. Descending into the bowels of the New York City subway system to await the D or the F or whatever train that would hurtle them to their destination. Waiting for the Fifth Avenue bus to get through the next stoplight and pick them up. Punching the elevator button, glad another day was over, and it was time to relax. Just another Tuesday.

It was a cloudless, crisp night, with a full moon coming up, its beams about to be lost in the brilliance of the blazing New York City lights.

Then it began. At 5:27 P.M. The unthinkable. New York City began going dark. Elevators stopped between floors. Subway cars stuffed with homebound commuters ground to a halt. Traffic lights went black.

New York City had lost its power. It was the Blackout of 1965, the largest electrical power failure in history. The city came to a virtual standstill as its citizens tried desperately to cope with civilization without electricity.

Almost a standstill, that is. There was one thing that never stopped working. One thing continued impervious to the bedlam erupting around it. One thing that kept a city in darkness in touch.

That one thing was the telephone. Friends and relatives across the country got reassuring calls from New York. A person might have been stuck on the eightieth floor of the Empire State Building, the only way of escape down an unlit stairwell, but that person was still able to call Kansas City and say, "I'm stranded but I'm okay." For when the mighty generators of Consolidated Edison flashed out, New York Telephone merely switched to its own stand-by power. Had the crisis gone on longer, it had other emergency power sources to backup the backups. All in all, New York Telephone statewide was able to handle a record 29 million calls in that crazy six-hour period.

AT&T's slogan used to be "The Bell System—it works!" It certainly did that day in 1965, because Ma Bell made it her business to be prepared—to give her customers service, with the lights on or off.

Today, when it's pointed out to a survivor of the '65 Blackout that the one thing that never stopped working was the phone system, the person is apt to look puzzled. There will be agreement, well, yes, the phones did work, but the puzzlement won't leave. The phones were expected to work. It was taken for granted—come hell, high water, hurricanes, tornadoes, blizzards, disasters of Man or Nature, Ma Bell could be relied on.

A crime has been committed.

The country has been robbed in one of the greatest ripoffs and dirty deals in modern industrial history. And in the commission of the crime, the nation's largest and most socially minded corporation was defiled and destroyed.

This was a crime that has cost us dearly:

- Residential telephone service has gone up some $200 a year on average, and is bound to go higher.
- More than $25 billion has been spent by the communications companies to comply with arbitrary rulings that had no benefit to the public.
- Technical advances and service improvements have been deliberately blocked by government actions.
- Our telephone service has lost its place as the best in the world, as other countries have surged ahead.
- Our balance of trade has suffered.
- Thousands of manufacturing jobs have gone overseas.
- National security has been damaged.

And there's more, so much more. Already the cost has been $800 billion and the final total just in terms of money could reach one trillion wasted dollars. The final total in terms of technology may be immeasurable.

This crime didn't happen all at once, in some shoot-'em-up, Wild West style. It was insidious, taking place over a number of years. It was the result of deliberately planned efforts of several government agencies and private businesses, creating the present situation out of a combination of ignorance and greed. This was a crime against society, against each and every one of us.

The country was happy with its Ma Bell. One poll even showed the highest favorable rating for any large U.S. company. Investors found it the safest money haven available and could count on the very nice dividends that were never skipped or reduced in the entire history of the company. Bell employees had well-paying and challenging jobs with lifetime security. Its scientists and engineers worked in an atmosphere that encouraged discovery and invention. And the telephone user? The telephone user received the world's best service at the world's lowest price.

So why was Ma Bell attacked? Who is to blame for the rape of Ma Bell? At whom can we point the finger? Take your pick.

A federal regulatory authority, usurping state powers, ignoring its mandate to protect the public's interest.

A Congress that refused to act.

A Supreme Court that wouldn't get involved in certain key issues, and made damaging, ill-informed rulings on others.

A group of equipment manufacturers and other entrepreneurs who saw a chance to make a killing.

Some government lawyers who saw a chance to make a reputation.

A press that failed to inform the public.

A Bell System management that was derelict in mounting an effective public information program.

A number of Bell System executives who were overly complacent and others who antagonized government officials by trying to block certain inevitable changes.

It's a long and ugly story, but to fully understand it we must start at the very beginning—with the birth of Bell's baby, the telephone. We will follow its growth, until, as Ma Bell, it became the world's largest corporation, a company with the principle of putting quality of service ahead of profit, one that nurtured technological and scientific advances, one that was finally cut down and destroyed because of a misconceived notion that big has to be bad.

In one sense, Ma Bell was sold out by her own people when they capitulated and gave in to divestiture. What followed has been a period of telecommunications anarchy during which the country as a whole and we as individuals have suffered.

But that need not be the end of the story. There is a future out there, one that can be better than what we have now.

This is a book about invention, about competence, and about integrity. It is a book about greed, about ignorance, and about deception. And it is a book about hope, about opportunity, and about challenge.

In the pages that follow we will present the story as we saw it from the inside. And we will make our case for what should happen in the years ahead.

Then, if New York City ever goes dark again, the telephones will stay lit, as they once did.

The Rape
of
Ma Bell

1

The Birth of Ma Bell

The baby that became Ma Bell was born on March 10, 1876, and its first words were "Mr. Watson, come here. I want you."

March 10, 1876, the telephone was invented. Alexander Graham Bell and Thomas Watson raced against time and strong competition to build a working model of an instrument capable of transmitting intelligible sound.

It was the stuff that movies were to be made of, but even the most imaginative screenwriter would have been hard pressed to come up with these lead characters. They were two men of divergent backgrounds brought together by luck and happenstance who through one great invention forged the course of a nation. Then they abandoned their baby to others and happily pursued different interests.

Only the wildest flight of fancy could have come up with an Alexander Graham Bell.

Born in Scotland in 1847, Bell showed little early promise, dropping out of school when he was 14. Casting about for something to do, he became interested in the problems of human speech and hearing. That was not too surprising since both his father and grandfather were prominent writers and lecturers in the field. Sometimes at lectures, his father called upon Bell and his brothers to demonstrate elocution and vocal sound production.

While Bell never went back to school, he studied anatomy and physiology on his own. Putting what he learned into practice, at the age of twenty-one Bell taught a deaf child to speak.

Shortly after this, both of his brothers died and the family moved to Canada. When he was twenty-five, Alexander Graham Bell moved to Boston to continue his work educating the deaf. Two of his young students were George Sanders, who was to become a lifelong friend, and Mabel Hubbard, whom he later married.

While working with the hearing impaired might have been personally satisfying and socially useful, it didn't take a genius to realize it wasn't going to earn Bell a lot of money. He wanted wealth. He wanted what his contemporaries, Thomas Edison for one, were acquiring. Inventions, yes, they were the ticket to financial freedom for a bright young man.

And Bell had an idea for an invention, something he called the harmonic telegraph. He was sure that musical tones could be used to send several telegraph signals at the same time thereby allowing one wire to do the work of many, saving money. Once he had the idea, he was obsessed—teaching the deaf during the day, experimenting at night.

The fathers of George and Mabel, Thomas Sanders and Gardiner Hubbard, were both wealthy and prominent businessmen. They became interested in Bell's experiments and financed them. But Bell needed more than money. Knowing little about electricity and mechanical construction, he needed technical expertise.

Then one day in 1874, quite by accident, he found it in the guise of a 20-year-old machine shop worker, one Thomas Watson. It was a match for the history books. Watson had a genius for mechanical construction, an extremely sensitive ear, and no fear of hard work. But best of all, he quickly developed an admiration for and devotion to Bell, who was seven years his senior.

Watson, by a strange coincidence, had also quit school when he was fourteen. At first he worked in his father's livery stable. Later he became a clothing salesman and then a machinist's apprentice in an electrical shop.

It was there that he discovered his true talent and interest.

He had a knack, a gift, for finding better and faster solutions to mechanical problems. If a job could be done more efficiently, Thomas Watson found the way.

This talent might be seen as an extension of his unceasing appetite for knowledge. In his spare time, he read not only any technical work he could find, but poetry, geology, and spiritualism as well.

The partnership began with Bell using Watson's shop to build models and parts for his harmonic telegraph. Then the men spent more and more time on the project together until finally they rented joint quarters where they worked and slept.

By June, 1875, the harmonic telegraph project was taking on a life of its own and pushing the men in a new direction—speech, sounds and words, not just tones, transmitted electrically. It was an exhilarating concept. Watson and Bell began concentrating on this more exciting project.

Bell said at that time, "I am like a man in a fog who is sure of his latitude and longitude. I know that I am close to the land for which I am bound and when the fog lifts I shall see it right before me."

Once he and Watson started down this new path, they found they were not alone. There were others working on the same project and the two men were in a race. However, Hubbard, their financial backer and Bell's future father-in-law, was not filled with the same enthusiasm. He had put his money up for the development of the harmonic telegraph, not this wild, science-fiction scheme. He threatened to cut off funds.

Fortunately he didn't, allowing Bell and Watson to feverishly create a working model of what was later dubbed a "telephone," an instrument that transmitted intelligible speech. "Mr. Watson, come here, I want you."

Ultimately this success would have meant little to Bell and Watson if the former, several days earlier, on February 14, 1876, hadn't gotten up early and hied himself down to the United States Patent Office and applied for a patent for the invention. Several hours later, one Elisha Gray, a Chicagoan widely known for his experiments in electricity, feeling he was far enough along in his work on the electrical transmis-

sion of voices, also applied for a patent. If Bell had overslept
that morning this book would have been about the Gray System and not Ma Bell.

The few hours made the difference. United States Patent
#174465, probably the most valuable patent ever issued, went
to Bell some seventeen days later, in what might be a record
for bureaucratic efficiency.

It took another few months before the American public was
let in on the secret. Bell's invention was demonstrated at the
Philadelphia Centennial Exposition by Emperor Dom Pedro
of Brazil. The emperor, marveling over what he heard, either
exclaimed "I hear! I hear!" or "My God. It talks!" The exact
words are in dispute since he marveled in Portuguese, a language not many of the bystanders in the City of Brotherly
Love understood.

Despite this success, money didn't come pouring in immediately. Strapped for cash towards the end of 1876, Bell and
Hubbard actually offered to sell the patent for a rumored
$100,000 to the Western Union Telegraph Company. Had it
gone through, it would have been on a par with Seward's
Folly and the purchase of Manhattan Island as a stupendous
buy. But Western Union, blindly confident of its supremacy in
the communications market, summarily turned them down.

Western Union was rueing its smug overconfidence within
a year. By the time it realized its mistake, the Bell patent was
no longer for sale.

Thwarted there, the large company tried to enter the telephony field through the side door. It sought out Elisha
Gray, the man who missed the patent by hours, thus becoming a curious historical footnote. Gray had already personally
conceded the patent to Bell. Besides missing the patent application race by a few hours, Gray had never actually built a
working telephone. He had the theory without the execution.

This did not faze Western Union. It acquired the right to
Gray's work and started up its own operation—the American
Speaking Telephone Company—late in 1877, to compete with
the Bell Company.

At first blush, Western Union had the advantage. For one
thing, it had an established network of lines and offices
across the country. Telephony was a logical extension of its

telegraph services. Western Union even hired Thomas Alva Edison to be its in-house inventor. It charged Edison with developing and improving the quality of the telephone. He promptly obliged with a new design for the telephone transmitter using the carbon variable resistance principle.

It was all-out war and chaos. City after city had two companies and two exchanges fighting against each other. Finally in September, 1878, the Bell Company sued on the grounds of patent infringement. The suit never made it to court. There was a settlement in 1879. Western Union faced the fact that Alexander Bell's beating Elisha Gray to the patent office gave the Bell Company the upper hand.

Western Union turned over its telephone business to Bell. In return it received twenty percent of Bell's telephone rental receipts for as long as the patent remained in effect.

A new company was formed to reflect the terms of the settlement. The combination of Bell Telephone and the Western Union phone interests became the American Bell Telephone Company.

Despite the resolution of the Western Union suit, the Bell Company had other challenges to face. Several inventors, including Phillip Reis, Amos Dolbear, and Daniel Drawbaugh, showed up with claims against the Bell patent. They had all experimented with the basic electrical-acoustical principles of telephony before or at the same time as Bell. However, like Gray, they had not produced a working instrument.

One by one their claims were defeated. Finally in 1888, the Supreme Court, by one vote, decided in favor of the Bell Company on the last case, Drawbaugh's, and all patent obstacles had been hurdled. After that, the American Bell Company's exclusive patent rights were never seriously challenged again.

And what of Watson and Bell? And Hubbard the financier? They had been pulled together for this one great endeavor and then all floated away, not to play any further significant role in the development and deployment of their achievement.

Within five years Hubbard retired. Alexander Graham Bell continued to teach the deaf, but didn't give up his inventing. He went on to experiment with the photophone, an artificial lung, a metal detector, and a hydrofoil boat. He researched the theory of flight.

He also set up the first Montessori schools in the United States and Canada. He was to become an advocate of equal rights and an outspoken opponent of racial discrimination. And as if that wasn't enough to keep him busy, he helped popularize the *National Geographic* magazine.

If Bell's interests were farflung and varied, Watson's were more so. He studied music, painting, paleontology, mathematics, and foreign languages. He took up farming. Then he became involved in education, building schools and heading school boards. Next his interest turned to building steam engines, devising a new bookkeeping system, shipbuiding, gold mining, and finally to playwrighting and Shakespearean acting. Throughout this diverse career, he never stopped studying, traveling, and exploring.

Bell and Watson were fated to come together that one time, to create history, and to move on along their very separate paths.

Despite being virtually abandoned by its parents, the fledgling company did remarkably well. A mere sixteen months after the first words over a telephone were spoken, the Bell Telephone Company was formed in Boston and the first commercial switchboard started up in Hartford, Connecticut. In May, 1877, there was a total of six telephones in service. Six months later, there were 3,000, and in three years, more than 100,000.

The first long distance call—albeit only eight miles long— was made in August, 1876, five months after Bell's invention. Eight years later commercial service was established between New York City and Boston.

It was a fast-growing company that needed a plan and a leader, someone with enough vision to tap its potential. In 1878 that man came to the Bell Telephone Company when Gardiner Hubbard hired Theodore N. Vail.

While Alexander Graham Bell gave birth to the telephone, it was Theodore N. Vail who fathered the Bell System. It was he who "invented" the system and gave it purpose. "One policy, one system, universal service," he was to say in 1910. That was the operating credo for what was to become the world's largest corporation.

Vail was born in 1845 and spent his early years in northern New Jersey. He got his first job at 19 working as a telegrapher for Western Union. A few years later he moved over to the U.S. Postal Service as a clerk. He quickly showed his executive talents and was named a superintendent of the Railway Mail Service when he was only thirty.

That was his position when Hubbard asked him to become the general manager at the new American Bell Telephone Company. This may have been the most significant personnel decision in Bell history. When Vail accepted, many of his friends thought he was crazy. Why forsake a promising—and secure—career with the Post Office for a new, risky company that was trying to sell, of all things, some silly gadget to the American public. It seemed sheer idiocy and professional suicide.

Besides, Vail didn't look to be, on paper at least, the man for the job. True, he was educated, well-read, and fluent in several languages, but he had little technical knowledge in the mechanical and electrical fields. These deficiencies didn't bother Hubbard and in the long run did little to diminish Vail's contribution to the Bell Company. What Vail lacked in the technical end, he made up for with managerial genius. He could deal as comfortably with heads of state as he could with factory workers.

And he was a man far ahead of his time. He staunchly argued that money was not the end, only the means. More important than making the greatest immediate profit was achieving the public's confidence. He also advocated employees owning stock in the company. These were concepts totally alien to most nineteenth-century businessmen and managers.

At first Vail's major concern was fighting the various patent challenges. Once that was out of the way, he was able to concentrate on structuring an organization that was to last and flourish, with little change, for a century until its forced dismantling under divestiture.

Vail reasoned that there were two essential elements needed to reach his goal of service that could be afforded by everyone. One was the creation of local telephone companies. These subsidiaries were nominally independent but were overseen, vigorously, by the parent company. Local companies, Vail reasoned, could better keep abreast of the needs of the local community. Thus were born the Illinois Bell, the Pennsylvania Bell, and the others.

The second element needed to provide the highest quality service was manufacturing capability. The Bell company should be able to set the standards and design and build its own equipment. Thus quality—and cost—control would be guaranteed. This arrangement would later be known as vertical integration and come under strong attack.

But to Vail it was logical and expedient. And so the Western Electric Manufacturing Company of Chicago was purchased from Bell's old nemesis, Western Union, in 1881.

By 1884 the basic structure of the Bell System was in place—a vertically integrated manufacturing and supply arm that put a strong emphasis on research and development and a network of partially or wholly owned subsidiary companies with strong central direction. This structure would last for a century without fundamental change.

In 1885, one further component was added to the Bell System—the American Telephone and Telegraph Company to handle long distance service. Vail became president of this subsidiary while remaining general manager of American Bell. Fourteen years later, the long-distance subsidiary merged with the parent company and the new enterprise took the name American Telephone and Telegraph, or AT&T.

Although Vail's contributions were monumental—as history was to prove—his long-term foresight did not sit well with the short-term expectations of the financial backers. They wanted profits and they wanted them right away. Rather than compromise himself, Vail resigned in 1887 after only nine years.

The period following Vail's departure was a difficult one for the young company. In 1894 the original Bell patents expired and there was a rush by other companies to enter the telephone business. Competition sprang up everywhere. In city after city, the consumer was offered the same service by different companies, service that was not interconnective. If a customer chose Bell, then that person could only talk to other Bell customers. Too bad if your mother lived around the corner and had a different telephone company. There was no talking to her by phone.

In many cases independent, non-Bell telephone companies were started in rural areas, not serviced by Bell. Naturally, this meant there was no duplication of service there. Many of these companies still operate today. But duplication, where it existed, made no sense. It translated into poorer service and higher rates. It cost far more to put in two, three, or more separate phone lines covering the same geographical area than it did to install one.

And duplication was clearly not in the best interest of the Bell Company. During the years 1890 to 1910, Bell bought up as many of its competitors as it could, and by any means at its disposal. One of its most powerful weapons was interconnection. Bell refused to let its competitors connect its local lines with Bell's long distance lines. Customers of independent companies could only make local calls within the companies' operating territory. If long distance service was wanted, one had to use Bell for local service as well. This eventually forced many of the independents to sell out.

As Vail saw it, "Two exchange systems in the same community, serving the same members, cannot be conceived as a permanency."

During this period there were impressive technological advances. The instrument itself was steadily improved. Away went the noisy single-wire-with-ground-return circuits in 1881 and in came the two-wire circuits.

In 1890 the first underground cables were put into place, ridding the city skyscape of the tangled web of wires that had threatened to keep the days forever overcast.

The loading coil was invented, increasing the distance intelligible sound could be sent. New York was connected with Boston in 1884, Chicago in 1892, and Omaha in 1897.

It was also in 1897 that Almon B. Strowger, a St. Louis undertaker, invented an automatic switch that opened the way for dial telephone service. This was one of the few significant advances in telecommunications made outside the Bell System.

Despite the advances and the success in takeovers, the

company remained troubled. Financially it was in bad shape. The price of AT&T stock fell 50 percent from 1902 to 1907. Its public image was dismal. People expected poor service from AT&T and they got it. The company had grown too fast and, with Vail gone, there had been little, if any, planning, leaving its employees totally demoralized.

In 1900 Vail was offered the AT&T presidency. He refused. In 1907, at 62 a respected elder statesman of American business, he was asked again. This time the challenge was too much. He accepted. In his next twelve years at AT&T, Vail charted the course for the future and put the company on it.

Vail found on his return that most of his first-term work had come undone. But he was firmly in charge and began a vigorous effort to get the company going forward again.

Vail again renewed his emphasis on service before profit. Further, he argued, the best service would be achieved only if telecommunications was recognized as a natural monopoly. Competition was wasteful and inappropriate.

As he put it, "The telephone system should be universal, interdependent, and intercommunicating, affording opportunity for any subscriber of any exchange to communicate with any other subscriber of any other exchange. . . . Some sort of a connection with the telephone system should be within reach of all. It is not believed that this can be accomplished by separately controlled systems . . . nor can there be competition in the accepted sense . . . this can be accomplished under such regulation as will afford the public much better service at less cost than any competition or government-owned monopoly."

Vail was not at all against public control of rates. He saw regulation as a natural outgrowth of a natural monopoly.

It must be remembered that in 1907, less than ten percent of the American population had telephones. To achieve his goal of universal service—which must have sounded incredibly optimistic and pie-in-the-sky then—Vail emphasized research, development, and long-range planning. He also pointed out the importance of uniform equipment. That would be achieved through the vertically integrated supplier, Western Electric.

It was Vail's belief that the corporation was responsible

equally to stockholders, customers, and employees. The Bell System's reputation for integrity and corporate ethics can be traced directly to him. He had no fear of open and public corporate dealings.

"We will lay our cards on the table; there is never anything to be gained by concealment. If we don't tell the truth about ourselves, someone else will."

This was no company honcho spouting hot air to obscure dirty dealings. Corporate honesty became ingrained at AT&T.

However, the mutterings of foul play got increasingly louder concerning AT&T's takeover techniques. Denying interconnection with long-distance lines came under more and more criticism. In 1913, the United States Department of Justice went so far as to suggest that these tactics might be in violation of the Sherman Antitrust Act.

Vail chose to compromise. AT&T would not buy any more independent telephone companies without prior government approval. And, most importantly, the remaining independents would be allowed to connect with the Bell long distance circuits. This agreement was known as the Kingsbury Commitment, named for the AT&T vice-president, Nathan Kingsbury, who laid out the compromise in a letter to the United States attorney general. The commitment effectively ended the threat of antitrust action and formed the basis for AT&T's relationship with the independent companies for the next seventy years.

The problem of employee dissatisfaction that Vail faced on his return was not so easily resolved. Twenty years had passed with no creative leadership and it showed.

Vail attacked by installing new staff heads, most notably John J. Carty as chief engineer. Carty presided over the vast technological progress achieved in the remaining Vail years. Many of the new appointees were examples of Vail's preference for broad-based generalists over telephony specialists. His preference was honored in the company for years to come.

Another boost to employee morale was Vail's hiring policy. He saw salary as the key. His employees would get the maximum wages in the field. No other company would pay higher than AT&T for comparable work. Years later this policy was gradually altered from the "maximum" wages paid by others to "comparable" wages.

Wages weren't the only employee concern before Vail's return. Personnel safety standards were so lacking that even the general public viewed telephone work as dangerous. Vail changed that. Also, in 1913, he instituted one of the first comprehensive industrial pension plans in the country's history. It served as a model for others ever since. His concern for his employees' welfare led to the establishment of convalescent homes for ailing workers. Again, in the last quarter of the twentieth century, these policies might be taken for granted. In the beginning of the century, they were truly radical.

Not unexpectedly, technology surged ahead during this era. Long distance calling was continually extended. By 1911 a New York caller—as long as he or she had a good set of lungs and could yell loud enough—was able to speak with someone as far away as Denver.

If service was to improve to where normal voice levels could be used, a breakthrough was needed—electronic amplification. Vail made this development a top priority. In 1913 the first vacuum tube amplification over telephone cables was tested.

This advance ushered in the entire electronic age. Harry D. Arnold, a Western Electric scientist, and Lee de Forest, an inventor outside the Bell System, share the credit for it. With this new tool, coast-to-coast telephony was inaugurated in January, 1915, when Alexander Graham Bell in New York spoke to Thomas Watson in San Francisco.

Later that year radio telephony removed the last distance barriers and an experimental call was placed between Honolulu and Paris.

Vail died in 1920. He more than left his mark—establishing the one policy, one system, universal service principle; ending wasteful competition with the independents; establishing the regulated monopoly principle for the telephone business; rescuing AT&T financially; emphasing the service objective as equally important with profit; making research and development a requirement for the long-term welfare of the company.

"Until Vail's day," Thomas Edison said in 1912, "the telephone was in the hands of men of little business capacity. Mr. Vail is a big man."

Vail is rightfully regarded as the founder of the Bell System, one that worked so well for so many years until outside forces with little understanding and no appreciation of it pulled it apart.

After Vail's retirement in 1919, Harry B. Thayer took over the presidency of AT&T. Thayer had not only been a vice-president under Vail, he was his close friend, sharing many of his views. Typical of the leaders that followed, the new president had spent his entire working career at AT&T.

It was Thayer's plan for the company to branch out into fields related to telephony. Congress had passed the Graham Act in 1921, formalizing the Kingsbury Commitment. Now AT&T appeared to be exempt from the Sherman Antitrust Act and had been recognized as a natural monopoly.

Thayer felt the company was in a secure enough position to expand. The natural arena seemed to be radio broadcasting, which owed its existence to a series of earlier Bell System inventions.

But before AT&T made many inroads into that industry, Thayer retired. His successor, Walter S. Gifford, took a different view. He wanted AT&T out of its nontelephone endeavors. He wanted the company's energies directed toward providing the best possible phone service at the lowest cost while still being financially sound and making a reasonable profit.

Gifford sold AT&T's radio interests to RCA, agreeing to furnish wire connections to radio broadcasters as required.

Gifford's tenure lasted from 1925 to 1948, longer than anyone else before or since. It was during his time that several significant financial events occurred. In 1921 AT&T established the famous $9-a-share dividend. Not even in the depths of the Depression, when earnings didn't fully cover it, was this dividend reduced.

It was during this period, too, that the general public began increasingly to buy corporate stock, "people's capitalism" it's been called. When Gifford retired in 1948, there were almost one million owners of AT&T stock—far more than any other company.

Oddly enough, while Gifford got AT&T out of radio, he got the company deeply involved in sound movies. As a consequence, until 1930 Western Electric had the talkies all to itself—it held the relevant patents and licenses. Gradually the company pulled back from the tinsel of Hollywood, and by 1935 was out of the movies as well.

This was to be a fertile time for technology. Commercial transatlantic telephone service started in 1927. In the same year, Harold S. Black of the Bell Telephone Laboratories developed the principle of negative feedback, which led to high-fidelity sound reproduction and electronic control of all sorts of mechanical devices. Nineteen thirty-six was the year of the coaxial cable, opening the way for television transmission and sending hundreds of calls on a single cable. The latter would greatly reduce the cost of long distance service.

Bell Laboratories also pioneered in experiments in television.

AT&T had every reason to feel confident and secure during this era. Even the Communications Act of 1934 seemed to consolidate its position. This act of Congress created the Federal Communications Commission (FCC) which was empowered as the regulatory authority over interstate aspects of the telephone business. The stated goal of the Communications Act was "to make available to all the people of the United States a rapid, efficient nationwide and worldwide . . . communication service with adequate facilities at reasonable charges."

This was interpreted as being supportive of Bell's use of subsidies to achieve high-quality universal service. These subsidies took various forms, such as business users supporting residential customers, high density routes helping out thin ones, and urban customers supporting rural ones.

The act also gave the FCC another responsibility. It was to see that no carrier constructed a new interstate line unless the commission first issued a certificate of public convenience and necessity. The intent was to prevent wasteful duplication of facilities and more capacity than necessary. This is a charge the FCC has virtually ignored since 1970.

The end of the Gifford years, in 1948, marks the end of the pre-modern era of telephony. The Bell System had reached its maturity.

2

The Golden Days

Once upon a time not long ago and very close by, there was a place where knowledge reigned supreme, where excellence was rewarded, and people worked for the good of others to provide the best service at the cheapest cost.

This place was not some faraway Camelot. It was right here in the United States, in our homes and offices. In our cities, suburbs, farms. It reached out and touched us all. And we took it for granted because it was so good.

It was the Bell System.

It is no exaggeration to claim that this organization, as it once existed, was a miracle of integrity and competence. Those traits were built into the system. They were demanded, expected, and they existed.

What was lost with the breakup of Bell is a national tragedy. It was a classic exercise in stupidity, greed, headline-grabbing careerism, and out-and-out thievery.

What was lost ranged from the most mundane—waiting longer to get a dial tone—to the most cosmic—the possible reduction of research of the planets, solar system, and universe.

Most people still do not comprehend the magnitude of what was perpetrated. To do so, a little time travel is in order, a step back to the halcyon days of the Bell System, from when it entered the modern age of communications in 1948 to its destruction in 1984. George Orwell would have chortled over the irony of that date. What follows about the old Bell might seem too good to be true, perhaps even distorted by rose-colored rememberings. But be assured, that's the way it was and the way it will never be again.

Theodore Vail, consummate planner that he was, could not have imagined in his wildest fancies what his baby would grow to be.

Before 1984, the company was the world's largest corporation, with more than $150 billion in assets and annual revenues of $70 billion—representing almost 2 percent of the United States's gross national product. Three million people owned stock in AT&T with no one person holding as much as 1 percent. It employed one million people and accounted for more than 5 percent of all corporate new construction. On a typical day the company handled a half a billion calls and made contact with forty million different customers.

The fact that something this large worked at all was amazing. The fact that it performed so efficiently and well was thanks to Vail and the structure he established. In essence what he did was divide the company into workable parts, creating quasi-independent units that were strongly interconnected and interdependent and that were overseen by 195 Broadway, the corporate headquarters in New York.

On paper the structure looked like this:

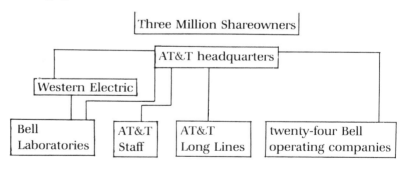

At the heart of the setup were the twenty-four local Bell operating companies, the New England Bell, the Illinois Bell, etc. They provided the local and intrastate long distance services. With the exception of the Southern New England Company (which covered most of Connecticut) and the Cincinnati and Suburban Telephone Company, AT&T was virtually the sole owner of the local Bell companies, the single stockholder. It was a substantial though not controlling stockholder in the other two.

Western Electric was the manufacturer, supplier, and in many cases installer of the equipment. Bell Laboratories was the research and development arm. Long Lines handled interstate and overseas calling. And the AT&T staff provided overall guidance and direction.

Uniformity helped the Bell System work and kept it from becoming too unwieldy. That Princess phone in Utah was the same as a Princess in Mississippi. The equipment in an office in Pasadena was interchangeable with that in Bangor. It was AT&T's job to insure this uniformity. Theodore Vail had said, "The network is one, and it must be designed, built, and operated as one."

In 1964, the then-president of AT&T, Frederick R. Kappel, said, "We must live in ... each city or village we serve and know and heed the interests and wants of local people. We are a big business, in short, that must be a small business, too."

The "small businesses" of AT&T were the local companies. That directory assistance operator who found the number of the corner grocery for you worked for the local company. The clerk who took your new service order—her paycheck came from the local Bell. The local company brought the behemoth of AT&T down to a manageable, human size. It was responsible for service in its franchised territory.

It was also up to the local companies to comply with all local laws and regulations. It's easy to forget that the FCC in Washington only regulated interstate business at that time. But each state had different rules and regulations and rate structures established by a public utilities commission.

To a great degree the local operating companies were autonomous although they had to stay within AT&T guidelines. Even if a local company experienced service problems or its earnings went down—because of nose dive in the economy or a state regulating board's getting contentious—generally the company took its own remedial actions.

As sole owner, AT&T, naturally, had every right to step in and make course corrections itself. That rarely happened. William S. Cashel, president of Bell of Pennsylvania in the mid-70s, once noted that the only direct order he ever received from 195 Broadway was to instruct him to strictly follow the rulings of the Equal Employment Opportunity Commission. It was an instruction sent to every operating company president.

Still, AT&T clearly ruled the local companies' boards of directors. Typically a board was comprised of business leaders, educators, and other leading community figures. There were approximately two hundred and fifty such members nationwide. One seat, however, was always reserved for an AT&T officer, usually a vice-president.

When it was time to select a new company president, the AT&T man brought a "suggestion" from New York. There's no record that this suggestion, coming from the major and often only stockholders, was ever ignored. The local president, therefore, was, in fact and in deed, part of the AT&T hierarchy.

As Cashel put it, "One aspect of AT&T control is certain: I was appointed to this position by 195 Broadway. Election by our own board of directors is just a formality. Our board gives advice, but seldom initiates action. An operating company president is in the Bell System corporate hierarchy as much as if the whole system were one integrated company."

Once installed, a president didn't receive a weekly checklist of things to do from New York. He had considerable freedom to run his company as he judged fit. If service results and earnings dropped, he—and back in those days, it was always a "he"—could expect a gentle reminder from 195 Broadway to take that remedial action if he hadn't done so already.

It took an extreme situation for a president to be removed from office—and even then he wasn't booted out on the street. Some nice "promotion" was dredged up for him—often a cushy staff position in New York.

One such extreme situation cropped up in the 1960s at Southwestern Bell when Ed ("Boom Boom") Clarke, a dynamic and ambitious man, was running the company. At that time, AT&T had recommended that the local companies install new, more modern central office equipment.

Boom Boom, who in truth was bucking for the top job at AT&T, wanted to establish his reputation back in New York by providing an impressive rate of return. He thought he could do this by overlooking the recommendation. Rather than replace the equipment, Southwestern Bell would modify it by adding electromechanical "brains," known as "senders" in telephonese.

Sure enough, it did save money and made the ledger tilt sweetly in the short run, but it also caused serious service problems and made future modernization more difficult.

AT&T headquarters let Clarke follow his misguided path— even his own engineers had objected to it. But when the problems began and snowballed, no one was terribly surprised when Clarke's career stalled.

Technically what the local company received from AT&T was "advice," and the advice was usually followed.

Such advice was part of the support services the AT&T staff and Bell Laboratories gave to both the local companies and the Long Lines Department. Innovation did not always originate at the Laboratories or with the staff. In reality, many innovations came from the local companies which were closer to the problems and service needs. It was AT&T's job to evaluate these proposals and determine if they should have systemwide application. The usual procedure was for AT&T to take one local company's idea and test it in another. If the test was successful, the idea was adopted.

Since the staff and the Laboratories did not generate income directly, their operating costs had to be picked up somewhere.

The somewhere was the local operating companies, and it was done through a fee (formerly called a "license contract payment") that was a fixed percentage of a company's revenues.

In the early 1970s each local company paid AT&T 2.5 percent of its revenues. This added up to a whopping quarter of a billion dollars annually. Many state regulating commissions, perhaps not understanding the services rendered by Bell Laboratories and the AT&T staff, found this arrangement objectionable, or, at best, questionable. They wanted to be convinced that the local customers—for it was the rates they were being charged that eventually ended up paying that license contract—were not being overcharged.

In 1972, AT&T and the local Bells did a thorough analysis of the AT&T staff and Bell Laboratories support functions to see if unnecessary and costly items were being charged to the local companies. At the same time they considered whether there were different and less controversial ways to pay for the services.

Several dozen senior managers spent three to four months meeting with every group and subgroup head at AT&T and Bell Laboratories. They went over with a fine-tooth comb what each group did and how it benefited the local companies.

In the end the arrangement was kept intact. By 1980, the local companies were paying $1.25 billion for the license service contracts.

This then was the corporate structure of the Bell System before 1984, the bare-bones skeleton. The flesh, blood, heart, and mind, were, of course, the people who worked for Bell, people who didn't so much join an organization as become part of a family.

3

A Modern Miracle

The system built by Vail was a modern miracle of integrity, competence, and devotion to service. From the point of view of the employees, it was a warm cacoon. Most Bell managers came straight from college and never left.

They were good jobs, coveted ones. At the height of its glory days, Bell employed more than a million people and a million more applied for jobs—every year! Only seventy thousand were hired annually with a mere four thousand going into management. That's only a 7 percent total turnover. When you worked for Bell, you worked for life—as is often the case in the highly praised and envied modern Japanese industry.

On average a manager stayed with the company thirty-two years, resulting in the lowest management turnover of any large American company at the time. Longevity was the norm. It was the expected. Recognition awards usually weren't given until 25 or 30 years of service.

Salary was not the number one reason employees stayed at Bell. Not that the salaries were shabby. The company had more or less continued Vail's stated policy of staying above average and somewhere near the top of the salary range for comparable jobs in other companies. While few got rich at Bell, few complained about the pay. In fact, in surveys taken in more recent years, employees griped about working conditions, attitude of supervisors, internal communications, and criteria for advancement, but pay complaints were seldom mentioned.

No, there was something else besides money that kept people at Bell for their entire working life. It's hard to explain to an outsider for it was not one, definite item that can be underlined and pointed to. There was an atmosphere at Bell, an attitude, where knowledge reigned supreme, where excellence and striving for the best product and the finest service was placed above all else—even the bottom line.

It was not the real world where dog-eat-dog competition was the norm. Excellence in performance was usually rewarded, and men with overriding ambitions such as Boom Boom Clarke would not get to the top. At Bell people pitched in and worked together to solve problems and raise the level of service even higher.

Working at Bell was a family affair, both literally and figuratively. Son followed father, daughter, mother, cousins and aunts and uncles. It was a place where early in his career the president of Pennsylvania Bell, Wilfred D. Gillen, was called Gilley and was proud he knew the first names of half the thirty-thousand-member staff.

Telephone company jobs were considered good by the community, and those who worked for the company were thought of as solid citizens.

They could take real satisfaction in what they were doing. Their job and goal was to provide the best service possible to the customer. The smallest decline in service, even if it would be virtually indiscernible to the telephone users and would save money, was not allowed. People like to take pride in what they are doing. And at Bell they could.

This dedication to customer service engendered an atmosphere within the company of mutual trust, respect, and integrity. It was an essential ingredient in the creation of what has been called "Bell's modern miracle."

For a company as large as AT&T to function efficiently and be the modern miracle it was, there had to be a well-defined organizational chart.

At the base of the operation were the nonmanagement or vocational employees. For many years the majority of these were operators. Interestingly enough, until recently they were women and they were predominantly Irish Roman Catholic.

It's speculated that women with a parochial school background fared better at this work because of the intense discipline nuns gave them. It took a lot of discipline to thrive in the rigid environment of the operator room.

Once management was reached, there was a ten-tiered progression from the densely populated level 1 of the supervisors and foremen to the resplendently lonely office of the AT&T chairman of the board up at level 10.

Level 9 was occupied by three or four AT&T officers with titles of vice-chairman or president. At level 8 were the executive vice-presidents of AT&T, and the presidents of Western Electric, the Bell Telephone Laboratories, and the local operating companies.

Into levels 6 and 7 fell the remaining officers of AT&T and its subsidiaries.

Level 5 was the lowest rung of "upper" management. In it were the department heads of the local companies—general managers and assistant vice-presidents—and Bell Laboratories and Western Electric executives with titles such as director or group head.

There were only some two thousand Bell System managers at level 5 and above. Just before divestiture the level 5 executive was earning approximately $100,000 annually.

The four remaining levels broke down to division heads at level 4, district heads at 3, and managers and supervisors at levels 2 and 1.

As one climbed the pyramid, job responsibility, naturally, increased along with salary. Perks and amenities got better as well. Levels 1 and 2 generally worked in open areas or if they did have an "office" it was not completely private. The walls were partitions and did not extend to the ceiling. Level 3 managers moved into private, if somewhat barren offices. There were no carpets, draperies, or easy chairs.

Reach level 4 and in went the wall-to-wall carpeting, draperies, and some place comfortable for a visitor to sit. By the time one reached level 5, status began to seep in. The office was larger. Besides the easy chair, there'd be a sofa, a secretary, and sometimes even a private washroom.

At level 6 and above, the impressive perks and amenities really began. The offices were plush, the limousines chauffered, and the long distance travel on company planes.

An individual's goal, in most cases, was to move up that pyramid. Built into the plan was an intense evaluation system giving the best people the best shot of doing so. In an organization as large as Bell, selecting leaders was no easy task and one that was taken very seriously.

Each year there were several thousand new management employees. Some were upgraded from vocational ranks. A very few came from jobs at other companies. The majority came directly from technical and liberal arts colleges. In this diverse corps were the leaders of the future, waiting to be identified, trained, and promoted.

The process by which this was done was in no way foolproof or perfect, but as it evolved over the years, it was effective.

Basically, in the first five years, a management employee was given difficult and challenging assignments to test the limits of his ability. (Again, until the seventies, there were few women.) A succession of bosses and management-development specialists evaluated his progress and performance. The employee was then rated numerically for his "ultimate potential"—that is, the highest organizational level he could reasonably be expected to reach. If the consensus evaluation was below 3, he was often advised to consider redirecting his career, probably outside the Bell System.

On the other hand, being rated level 5 or higher meant the employee was on the fast track and received special attention. He was tagged for bigger and better things and got additional training, transfers to broaden his knowledge and viewpoints—and more frequent and intensive evaluations.

A successful college graduate, hired at level 1, could generally expect to reach level 3 within 10 years. Most stayed at that level for the rest of their careers.

Good talent sometimes fell between the cracks even with this intense scrutinization. A capable young manager could have a boss who just didn't like him or didn't have the skill to talk him up at the periodic evaluation meetings. The young manager's career might then be stalled. On the other hand, the process pretty well kept less-talented managers from rising rapidly through the system merely by playing politics. One didn't move upward without a consensus that the individual had what it took, that he had the right stuff. It was rare that a particular manager got ahead because someone was pulling strings for him. There were just too many different people making the evaluations.

In addition, an employee didn't stay in one place. Inching up the Bell pyramid entailed a good deal of moving, both organizational and geographical. At the lower levels movement was between departments or divisions within a department. It also involved moves between line and staff jobs.

At higher levels, transfers were common between areas and even companies. In all cases, there were two primary objectives. One was to give the employee additional experience and a greater knowledge of the business. The second was to expose him to a new group of senior managers who would then evaluate him.

In large urban areas, particularly New York, a Bell System manager could make a large number of interorganizational and even intercompany moves without having to relocate his home or family. Outside of densely populated areas, employees were expected to uproot their families every few years. Moving came with the territory if one wanted to prosper at Bell. It wasn't unusual for a manager to buy and sell a half a dozen or more homes in the course of an upwardly mobile career.

The pace of this uprooting and moving slowed down in the seventies and eighties. It is unclear exactly why. Perhaps it was felt that so much crosstraining was unnecessary. Or it may have been an adverse reaction by employees to the strain on the family or the lack of community identification. These employees had become nomads who couldn't say they were "from" anywhere—except Ma Bell. Of course, employees had such a strong "family" identification with Bell, that this was not as onerous as it might have been at other companies.

Another factor for the transfer slowdown may have been the soaring costs of moving, which the company paid. By the early 1980s, to move a manager and his family cost between $50,000 and $100,000. The company footed the bill for more than the moving van. It reimbursed the manager for the costs of buying and selling houses and made up the difference if the new mortgage interest rate was higher than the last.

Bell people worked and socialized with each other. There were company picnics, group trips to Europe and other vacation spots. Bell also had an organization unique in American industry. It was called the Telephone Pioneers of America. Membership was open to employees with twenty-one years or more service and to retirees. The Pioneers were both a social group and a community service organization. They had parties and annual conventions. They worked in hospitals, with the blind and deaf, and elsewhere. While the Bell System provided some subsidy, it did not supervise or direct the group's activities. The Pioneers served to bind together long-service employees in off-the-job situations.

There was never any slowdown in training, which could also set an employee to packing his suitcase. Managers continually received technical and managerial training—and sometimes even took courses in the arts. Someone at the company for forty years might spend as much as seventy weeks in fulltime company-sponsored courses. While a manager might attend classes at outside colleges, the Bell System also ran its own university at Lisle, near Chicago. In addition to company-run programs, employees were encouraged and subsidized in out-of-hours education they arranged themselves.

Wherever possible evaluation was done by the numbers. If performance could be given a number, this was done and it was reported frequently.

Every employee, from the chairman on down, was subject to this numerical measurement. On the lower levels, managers were rated on costs, service, and productivity. At the higher levels, the ratings included corporate financial performance such as budgets, revenues, and earnings.

Typically a manager began each year by listing his performance objectives for the next twelve months, his New Year's business resolutions. After review and approval by his boss, these objectives became his key motivators. Periodically (usually monthly or quarterly) the manager reviewed progress with this boss and adjusted his actions, if need be, to better reach his goals.

Objectives were seldom changed during a year. They were firm commitments. One's success as a manager was measured by the extent to which these commitments were met or exceeded.

Numerical measurements not only determined a manager's progress towards his yearly goals, they fostered fierce internal competition between managers. In the automotive field, Ford fought it out with General Motors. In telecommunications, AT&T managers did not get complacent because there was no real outside competition. They worked harder because of the internal competition.

Managers with similar responsibilities across the country continually compared their service and productivity results with each other, sometimes even making friendly cash wagers on the side. An enormous effort went into designing appropriate systemwide measurements.

"If it isn't measured," Robert Gryb of the AT&T Internal Measurements Group once said, "it doesn't get done. If it's not measured right, it doesn't get done right."

These measurements appeared in dozens of publications, the most famous being the monthly Green Book. It contained more than forty pages of charts and tables, recording everything from job injuries to unkept installation appointments.

Each local company and each area within the companies (there were 95 areas nationwide) were compared and ranked in dozens of categories. An area was usually headed by a Level 6 manager who seized the Green Book the instant it came in to see how his area ranked in the overall summary and to see if any performance category had slipped.

The level 6 manager wasn't the only one scrambling to get the Green Book. His subordinates were anxious to see it before he did, presumably to get their excuses ready.

Each manager wanted his ranking to be high. Any officer whose area was consistently near the bottom of the charts knew he'd better improve those numbers if he wanted to get promoted or even keep his present job.

There were scores on just about everything including:

Percent of call attempts receiving a dial tone within 3 seconds.

Percent of repair requests serviced within four hours of receipt.

Average time to handle a directory assistance request.

Average speed of operator answer when a customer dials "O."

Average holding time for a customer's business office call.

Central office lines not functioning.

Subscriber cable pairs not functioning.

Percent of long distance calls blocked because of insufficient trunks.

Central office maintenance dollars spent per lines served.

Outside plant investment per subscriber line.

Customer billing errors.

Installation service quality.

Transmission quality.

These measurements and combinations of them made it possible for a single number to express the overall quality of local dialing, repair, and installation services along with many other activities. All of these were published quarterly, monthly, and in a few cases, even weekly. Management wanted to keep on top of operations and management did.

These numbers didn't pop out of thin air. There were various methods used to come up with them. In some cases, direct mechanical or electrical measurements were used. Others were based on sampling, with only some of the results recorded. Then there were human checkers, called service observers, whose identities and activities were kept as secret as possible from other employees. Service observers monitored actual operations and collected and recorded data of all types.

Some measurements were made, or at least validated, by middle managers on a random basis. Usually a manager didn't evaluate his own department; instead he observed another to avoid conflict of interest. As a last resort, self-ratings were used. Few employee activities escaped service measurement.

To have a uniform basis of comparison, everything was rated on a scale from 0 to 100.

While 100 was the maximum possible score, it wasn't necessarily the best one. Generally a rating of 99 or 100 meant service was too good, that money was being spent beyond the point of effective return. (Very few managers, however, were reprimanded for 99s or 100s.)

Scores of 96 to 98 were the target goals and managers strived to keep their scores within that range. Even dropping as little as two points to 94 was considered a serious problem. Below 90 was clearly unsatisfactory. These were labeled "weakspots" though in truth they could be career-threatening crises. It was believed that performances below 90 were noticeable to the public and led to customer complaints.

There were dangers in rating by the numbers. One was cheating through inaccurate reporting of data. Cheating was not only considered a serious offense, it was also viewed as unsportsmanlike, somewhat akin to taking a few strokes off a golf score. But cheating occurred very rarely. The atmosphere of ethical conduct was not conducive to it.

Another danger of numerical rating was that performance indicators can't measure everything or make judgment calls. And occasionally the desire to provide excellent service and also demonstrate high efficiency led to overdoing both jobs. An example of that occurred in 1963. It was an isolated incident and, once discovered, immediate steps were taken by management to correct it.

At that time a major performance measurement for the central office equipment was "dial tone speed." This was the percentage of customer calls that didn't get a dial tone within three seconds of picking up the handset. The objective was to keep this at 1.5 percent. A higher figure meant a shortage of central office equipment and inadequate service. A lower figure suggested that too much central office equipment had been installed and therefore more money had been spent than necessary.

During minor economic slowdowns, fewer calls are made. Customers cut back to save money. This meant that more callers received a dial tone within three seconds because the central office equipment was not being used as much. To keep the dial tone speed numbers looking good, "line finders"—a piece of equipment relating to dial tone speed— were being removed and junked. Of course, when the local economy picked up, line finders had to be reinstalled.

Research showed that in one central office, line finders had been removed, installed, removed again, and reinstalled almost every year for eight years, keeping the dial tone speed at a "perfect" 1.5 percent. Had the measurement been allowed to fluctuate from 1.5 down to .5 percent, the customers would have had better than normal service and the cost of the installation and removal would have been saved. It was an example of the rating system being misused and the numbers being the end and not the means to the end.

For the most part, the index system worked well. It focussed attention on service and cost. The Green Book and the great many other publications in more specialized areas fostered the company's internal rivalry. This rivalry made the Bell System special, and helped further the quality-of-service attitude that reigned at the company. The last issue of the Green Book appeared at the end of 1983. It was another casualty of divestiture.

Any drawbacks of the Green Book and other rating systems were minor when compared to their overall value. Problems got identified and dealt with quickly. For instance, in March, 1961, a service problem reached crisis status in Philadelphia. It had to do with something called the Billing Service Index (BSI), which measured the timeliness and accuracy of a customer's bill. This index had many components and was more difficult to understand and analyze than most.

Billing was one of the more complex activities to measure since many departments were involved. A telephone operator had to prepare toll tickets or key calling-number information into the computer. Automatic message recording equipment had to work properly. The accounting department had to process the data and prepare the bills. If there was a billing complaint, the service representative had to handle the problem correctly. Technicians had to keep switching equipment well-maintained. All these were factors in getting a correct bill to the customer on time.

In 1961, Philadelphia was performing absymally by Bell standards. Its composite billing service index had been on a gradual slide for six months before dropping to 67. This made it the second poorest in the Bell System.

The president of the Pennsylvania company demanded immediate remedial action from the Philadelphia Area general manager. The GM was faced with that classic management problem of where does the buck stop. It's the tendency when several departments are involved in a problem to pass the blame to someone else and expect another department to fix it. The general manager chose a course often used at Bell. He brought in someone from a different Bell area to chair a task force of managers from each department involved in billing.

The group spent three months analyzing the situation full time. They identified the problem sources and then examined them in detail. They visited other Bell offices where such problem areas functioned well. They compared other procedures to those in Philadelphia. Then they made their recommendations for changes in a written report.

The Philadelphia general manager okayed the report in July and the recommended changes were put into effect immediately.

Even before the report was written, there had been improvement. The BSI climbed to a more satisfactory level of 80. Bell managers had long since noticed that simply focusing attention on a problem generates improvement. People tend to work harder and more effectively when they are being watched.

Six months after the recommendations were implemented the BSI was at a nearly satisfactory 93.

This was not an isolated incident. The greater the problem, the more the resources of the system were mustered to correct it. In 1969-70 when local service in New York City suffered a severe and well-publicized breakdown, a huge force from the entire Bell System was mobilized to bring service back to a satisfactory level.

Part of what made the Bell System so successful was its emphasis on long-term planning. Few if any business organizations planned with the intensity and dedication of Bell, starting with its original and ultimate planner, Theodore Vail.

Furthermore, there were few other companies whose planning horizons were as distant as Bell's. For example, as soon as the first electronic switching system was placed in service in 1965, a target date of 2000 was set for the completion of the electronic conversion of all central offices. (It now looks as if that goal will be met a decade ahead of schedule.) Targets were set for every sort of technological innovation in the network. And they were not idle speculations over a couple of beers. No, the targets became the basis for budgets, personnel planning, and manufacturing. Very few goals were ever missed.

In the technical area, detailed planning is obviously needed for bringing a new installation on line. For instance, let's say economic studies have shown a new central office location is needed. It will take about four years from the time the decision is made to when it's opened. Land will have to be acquired, the building designed and built, equipment manufactured and installed. And finally there will have to be pre-service connection and testing. All during this time, plans must be continually reviewed to consider the effects of any changes in the basic assumptions.

Projects were carefully scheduled so that all related parts—central office equipment and outside plant, for example—were completed at the same time, otherwise large capital investments would sit idle without earning a return. Discrepancies of a few weeks were considered unacceptable and would subject an engineer or manager to severe criticism. Compare this with how our public highways are constructed, with completed segments sitting unused and deteriorating while connecting sections remain unbuilt.

Long-range plans were also prepared and published for financial targets. Usually they covered a time span of five or six years and required information from all departments of the company. These plans took into account expected growth, business conditions, regulatory climate, price inflation projections, available technology, and so on. These documents served as valuable guides to executive decision-making. The widespread availability of computers in recent years made it possible for managers to inject all sorts of "what if" scenarios into these business plans and to then assess them.

But, of course, long-range planning does not work well with major discontinuities—such as those introduced by the FCC and by divestiture. But even then, once the facts were established, corporate planners worked overtime to include the effects of these drastic changes in their planning.

Service was the goal whether it was in long-term planning, short-term programming and installation of plant additions, or day-to-day operation and maintenance of the network. The highest quality service and performance was uppermost in the minds of Bell managers and craftspeople. Whenever a choice had to be made between cost and service, or between revenues and service, service considerations invariably won out.

Someone might argue that it is easy for a monopoly, such as Bell was, to give quality service. The company could bury the costs of free repair visits and the almost instantaneous response of operators into overhead. Get the public utilities commissions to raise rates to pay for the service. But good service and monopolies don't necessarily go hand in hand. Just ask anyone who has suffered the agony of dealing with a bad utility company. Ten hours in a dentist's chair without Novocain is preferable.

Again, if it were a question of profits versus service, profits lost. For years the California regulatory commission held the rates there so low that the local operating company could not have survived by itself maintaining the AT&T standard of service. Another organization might have decided to balance the books in California by cutting back service there. Not AT&T. The rest of the system subsidized the California company. Bell had the single-minded aim that nothing—not even California utilities commissioners—would interfere with giving every individual customer the best there was to offer.

"... the priorities are clear," AT&T chairman Charles Brown stated in 1980. "Our objective is to render high quality and constantly improving service. A competitive return on invested assets is a *sine qua non* of that objective. But let's not get the two reversed."

This policy did come under attack. Judge John F. Grady,

sitting on the case of MCI vs. AT&T in 1980, severely criticized
former AT&T chairman John deButts for putting customer interests ahead of stockholders'. On the other hand, two years
later Judge Charles R. Richey was to take the opposite position, commending instead of condemning Bell for making
quality service its primary goal.

This dedication to service engendered an atmosphere
within the company of mutual trust, respect, and integrity.
Beyond the doctrine of service, Bell employees knew they had
to respect the privacy of the customer's communications. A
conversation overheard was a conversation forgotten.

This was no mere custom. This was a written rule set down
in a document called "A Personal Responsibility." This book-
let spelled out a code of conduct concerning company funds,
fair competition, proprietary information, and conflict of interest, among other items. If someone wanted to be fired, he
had only to violate the communications privacy rule, and he
was out.

There was one case where an employee suspected his wife
was fooling around. To get some evidence, he put a tap on his
own phone to monitor her calls, reasoning the company rule
didn't apply in this instance. Wrong. He was fired and prose-
cuted. Communications privacy was sacrosanct. Period.

AT&T chairman John deButts said in 1975, "The character
of the Bell System—its reputation in the public's mind—is
quite literally in the hands of tens of thousands of people
who make up management. There is no place so remote and
no manager so obscure as to not offer the potential for
influencing—perhaps profoundly—the reputation of the Bell
System as a whole. This is our greatest vulnerability. It is also
our greatest strength."

There was, in fact, a high standard of business ethics. Gifts
from contractors, no matter how small or trivial, were strictly
forbidden. There was no commingling of company and personal expenditures. Cheating on expense accounts was virtually nonexistent.

But as deButts suggested, any organization with people is
vulnerable and it's to be expected that impropriety—and
worse—sometimes, though very seldom, raised its ugly head
at Bell.

On a minor level, there was an incident in western Pennsylvania in the 1960s. It began mundanely enough, but quickly made its way into AT&T folklore. One day the plumbing backed up in one of the company's operator offices. It was soon discovered the blockage was being caused by a large number of used condoms clogging the drain. How they got there was something of a puzzle since only women used the building.

But it didn't take Sherlock Holmes to solve the mystery. It seemed that the night chief operator had gotten a tad carried away in providing "service" to the local male population. Showing a certain amount of entrepreneurial flair, she had set up shop in the employee lounge where a few of her subordinates gave a lot more than directory assistance. The chief operator was later disciplined for letting unauthorized people into the building.

What happened in Pennsylvania in 1982 was taken a bit more seriously. At that time the state public utility commission was considering a telephone rate increase request when one of the commissioners had his picture taken by a newspaper photographer at a Philadelphia country club. Unfortunately, the commissioner happened to be the guest of several top Bell of Pennsylvania executives. Much was made in the papers of the "attempt to influence the regulatory process." The incident surely showed poor judgment by the Bell executives, but no one should have seriously believed the rate case would be affected by a few holes of golf. Still, it was a breach in company policy.

What was taking place in Texas at about the same time was far more serious.

It began with an internal investigation by Southwestern Bell of one of its more flamboyant executives, T.O. Gravitt. The company wanted to know if the fifty-one-year-old vice-president had improperly diverted some company funds for his own use. Apparently Gravitt was unwilling to see the investigation through to its end. He left his car engine running in a closed garage, committing suicide on October 17, 1974.

That tragedy might have closed the case except for the note—and verifying papers—Gravitt left. In it he charged Southwestern Bell with a catalog of offenses from political payoffs to illegal wiretapping. What caused the biggest flap, however, was the claim that the phone company had gotten rate increases from the state by falsifying information.

"Watergate is a gnat compared to the Bell System," Gravitt wrote.

It was a terrible set of accusations and it had 195 Broadway rolling with the aftershocks. AT&T conducted an audit of each of its local companies' activities. But beyond what had happened in Texas and North Carolina, only minor improprieties were uncovered, and those mostly concerning misuse of funds.

All this led Chairman John deButts to say, "Compromise of the integrity of this business casts a shadow over everything we do and impairs our ability to reach every goal we strive for. . . . We are determined to assure that nowhere in our business today is there any trace of the irregularities in the conduct of our affairs and regulatory activities that have been attributed to us and, if there is, that it will be rooted out unequivocally and forever."

These examples are cited because they represented exceptional behavior. The revelations were shocking just because they were rare.

There was a painting that hung in many a Bell manager's office. By artist Frank Merritt, it showed a lineman at the end of the last century trudging on snowshoes through a blinding blizzard looking for a break in a toll line. The painting was called "The Spirit of Service."

There was a spirit of service in the old Bell System, a spirit indoctrinated in each employee from day one. It wasn't a slick corporate slogan either, it was a way of life for Bell employees.

The United States now has a so-called service economy. While "quality of service" is boasted of by many companies, it is delivered by few. Try to find a clerk in most department stores who is familiar with the stock and isn't gossiping about last night's date. Bank tellers close their windows at coffee break time even if the waiting crowd has just reached a peak. And too bad for any airline passenger with the temerity to ask for a pillow when the stewardess is serving drinks.

Tell your children of the days when doctors made house

calls, stores made free deliveries, and the milkman dropped off your order every day, and they'll look at you as if you're talking about the ancient Roman Empire.

Economic factors such as price competition and labor shortages partially explain away the service problem. And public attitude does as well. We've come to expect to serve ourselves at gas stations, discount stores, and automatic money machines. And we are profoundly grateful when carpenters and electricians actually show up, as if they are doing the customer a favor.

A funny thing happened since divestiture. There's a serious tear in that "Spirit of Service" painting and it's unlikely it can ever be fully restored.

4

Ma Bell's House of Wonders

"The voice of the Bell Labs was like the voice of God."
—AKIO MORITA, Chairman of Sony Corporation

It was a national treasure nonpareil. It was the greatest industrial research laboratory the world has ever known. It produced more new technologies than any other research laboratory on earth. Out of it came a chess-playing computer that could challenge grand masters as well as convincing evidence supporting the Big Bang Theory of the creation of the universe.

It was the Bell Laboratories, Ma Bell's House of Wonders.

It all began back in January, 1925, when the Laboratories were founded as an offshoot of Western Electric, which since 1907 had been handling the Bell System's research and development work in addition to manufacturing and supply. Since then few advancements in telecommunications have been originated or brought to fruition outside the Laboratories. One reason so much was accomplished there was its mandate to design and develop everything connected with telephone service from one end of a connection to the other, and to make it all fit together well. Of course this responsibility has been divided since divestiture and there is no doubt that the Laboratories' primary mission has changed and as a result, this national asset has been severely damaged.

Before the 1984 breakup of AT&T, the Bell Telephone Laboratories had some twenty-five thousand employees at nineteen locations and operated on an annual budget of more than $2 billion. There were three thousand Ph.D.s working there, more than at any other organization, including universities.

Through the years Bell Laboratories scientists and engineers obtained twenty thousand patents and won ten Nobel prizes. Their work ran far beyond telecommunications, as they engaged in pure research in mathematics, chemistry, psychology, even into DNA molecules.

Only a large, well-funded organization could afford to give scientists the freedom they enjoyed at the Laboratories. Many of their activities led nowhere—that's the nature of research—but others had surprising results, such as the foundation of radio astronomy by Karl Jansky in 1933, and Arno Penzias and Robert Wilson's confirmation of the Big Bang Theory in 1965. What made Bell Laboratories remarkable was at one end of its spectrum there were people free to dream and at the other practical engineers influencing the day-to-day operations of the Bell System.

And the dreamers were free to dream. A scientist, giving a demonstration of a device which accurately synthesized the sound of a violin, was asked what was the purpose of his work. Purpose? The scientist was puzzled. It had never occurred to him there should be a purpose.

And while a mechanical mouse threading its way through a maze by trial and error, and then learning from its past mistakes on the next go-through, has a relationship to computer artificial intelligence, it was built to satisfy scientific curiosity and not for a specific goal.

The Laboratories attracted a special breed of people. If there wasn't enough time in the regular workday, they found time elsewhere. Research and knowledge was the thing. Robert R. Williams had something he was working on, but he didn't want to take away from what he was doing during the day as director of chemistry for the Laboratories. So for twenty years, he worked after hours in his garage until finally in 1934 he managed to synthesize Vitamin B1 from rice hulls, bags of which had been crammed into his garage for all that time.

The complete description of the significant scientific achievements of the Bell Laboratories fills seven volumes and five thousand pages in a series entitled *A History of Engineering and Science in the Bell System*, but here are a few highlights.

- The first cathode ray tube, a predecessor of the television tube, was made practical by J.B. Johnson in 1923.

- Harald Friis developed the first superheterodyne radio receiver in 1925, working on the principle that received radio signals could be better amplified if they were converted to a lower frequency. Today all amplification of radio signals, including television, radar, and control signals, use this principle.

- In 1926 H.M. Stoller and A.S. Pfannstiehl invented the first machine to synchronize motion picture sound, bringing on the age of the talkies.

- Herbert Ives and Frank Gray developed the first long distance television transmitting and receiving equipment, publicly demonstrating it in April, 1927, with a transmission from Secretary of Commerce Herbert Hoover in Washington to AT&T President Walter Gifford in New York. The Laboratories demonstrated color television two years later.

- Clinton Davisson won a Nobel prize for his work in 1927 with Lester Germer that showed electrons projected as particles also act as waves and obey the laws governing radiation.

- While riding a ferryboat across the Hudson River on August 2, 1927, Harold S. Black came up with the principle of negative feedback—inverting and returning a small portion of the signal output from a vacuum tube amplifier to the input reduces distortion and improves quality. This made possible later advances in telecommunications transmission and high-fidelity sound reproduction. While Black had the idea, it was Harry Nyquist and Hendrik W. Bode who actually produced a working amplifier based on this principle.

- Dissatisfied with the poor sound quality of radios and sound systems, Harvey Fletcher produced a stereophonic reproduction system in 1930. It was first demonstrated publicly in 1933 with a broadcast of a Philadelphia Orchestra performance from Philadelphia to Washington, D.C. Famed conductor Leopold Stokowski handed his baton over to an assistant so he could personally take charge of the sound system controls.

- Lloyd Espenshied and H.A. Affel in 1930 developed a coaxial cable carrier system that provided a very wide bandwidth capable of carrying thousands of telephone channels or several video signals.

- W.M. Bacon originated the automatic store and forward switching concept in 1934. This provided fast and efficient handling of teletypewriter message traffic. Although superseded by computers, the basic principles are embodied in today's packet switching systems.

- Using electromechanical relays, G.R. Stibitz constructed the first digital computer in 1939.

- In 1940 Clarence Lovell designed the first electronically controlled gunfire control system, the M-9 gun director. The idea came from a dream one of Lovell's co-workers, David Parkinson, had. Parkinson dreamt that radar could not only detect incoming aircraft but could automatically aim anti-aircraft weapons as well. Lovell didn't think this sounded far-fetched at all, and went on to design the M-9. The system succeeded in destroying 76 percent of the German V-1 buzz bombs aimed at London late in World

War II. The M-9 and other systems that followed evolved into control arrangements for the Nike and several later series of missiles.

• Harald Friis and A.C. Beck led the development of micro-wave radio transmission in 1946 using J.A. Morton's invention, the close-spaced triode, a vacuum tube micro-wave amplifier. Until the availability of optical fibers some thirty-five years later, microwave radio was the dominant long distance communications facility.

• William Shockley, John Bardeen, and Walter Brattain won the Nobel prize for their invention of the transistor in 1948, probably the most significant development in the field of electronics in this century. It is the basis of the entire computer industry.

• In 1954 Sidney Darlington conceived the basic principles of the Bell Laboratories Command Guidance System, which was subsequently used in most of the free world's scientific space vehicles. There have been no failures of this system in use.

• In 1954 D.M. Chapin, C.S. Fuller, and G.L. Pearson invented the solar battery, the first efficient means of directly converting a significant amount of the sun's energy into electricity.

• In 1958, Charles Townes and Arthur Shawlow invented the LASER, or Light Amplification by Stimulated Emission of Radiation, which has revolutionized telecommunications and found numerous applications in medicine, industry, and warfare.

• P.W. Anderson received a Nobel prize for his 1958 theoretical work in physics dealing with the nature of electrical conduction and magnetism.

• John R. Pierce conceived the idea of communications satellites in 1954, several years before artificial satellites of any sort had been launched. His scheme was first tested by the Echo satellite in 1960.

- Andrew Bobeck invented magnetic bubble technology in 1966. It provides for storage of data on magnetic material and has application in large-scale memories such as those used in computers.

- In 1969, Jack Morton decided it should be possible to apply the principles of Bobeck's magnetic bubble technology to semiconductors, using electric charges in silicon to perform memory and logic functions. He called two Laboratories' scientists, Willard Boyle and George Smith, into his office and suggested they get to work on it. Within one hour, they had sketched the basic ideas that led to the Charge Coupled Device, or CCD, forming the basis of a new industry. The CCD introduced a new concept of photoreconnaissance through visual imaging. With this, images are relayed digitally from communications satellites for computer manipulation and retrieval. Current spy-in-the-sky techniques are based on the use of CCDs.

- Work at the Bell Laboratories in glass technology made optical fiber transmission practical through the development of extremely transparent glass fibers. Optical fiber transmission is now being used in nearly all new long distance facility installations.

New inventions were not the only revolutionary things to come out of the Laboratories over the years. Bell scientists and engineers also invented new methods of applying technology, some of which spawned whole new industries. These include telephone traffic engineering, quality assurance, information theory, stored program control of switching systems, and systems engineering.

While a few Bell Laboratories scientists were given free rein to work on favorite projects, most research and development work was more purposefully organized. The idea for a project might originate anywhere, but at some point, the appropriate laboratory group had to assemble a case. A council of AT&T, Bell Laboratories, and Western Electric managers would then review the case to determine if it were practical and economical. If the answers were "yes," the project was approved and funded. After that, periodic progress reports were given to justify continuing the project, and sometimes to get increased funding.

There were fewer personnel transfers between the local Bell operating companies and the Laboratories than among other parts of the Bell System. Laboratories jobs required specialized advanced educations and degrees not generally held by people with the local companies. However, there were interchanges involving administrative personnel, and also with people in systems engineering. Local company managers could provide valuable knowledge of field needs and practicalities. Many innovations came from operating companies where the need for improved service and efficiency was most noticeable.

The Laboratories did have close working arrangements with the local companies. Sometimes local company managers took part in work-study programs at the Laboratories aimed at raising their technical qualifications.

There was a continuing concern at the Bell Telephone Laboratories that it keep its feet on the ground. In addition to advancing the frontiers of science, it had to remain aware of the mundane day-to-day needs of the local companies. For example, redesigning a gasket for the watertight door of an equipment cabinet mounted on a pole can hardly be considered state-of-the-art, potential Nobel-prize-winning technology—but it was necessary technology. So was designing a plastic cable sheathing that squirrels and rats don't fancy for their dinner.

Bell Laboratories maintained field representatives at each local company to help sustain a good two-way flow of communication. Bell operating company engineers met with these people frequently. The individual representatives remained on their assignment for only a few years before rotating back to other Laboratories jobs. This insured their remaining conversant with current Laboratories activities and capabilities. Another function of the representatives was to determine the quality of Western Electric products used by the local companies and to institute corrective measures when necessary.

In December, 1983, just before divestiture, these field reps locked up their offices for the last time, said their goodbyes, and went home to the Laboratories. It had been decreed that the local companies and the Laboratories would sever all corporate ties and end their close working relationship. Another valuable resource had fallen victim to divestiture.

Bell Laboratories survives divestiture as a division of AT&T with the official new name of AT&T-Bell Laboratories. It is still staffed by top-notch scientists and engineers, and there is little doubt that its long list of achievements will continue into the future. Work is continuing on the photon revolution, in molecular electronics, computer science, and other advanced fields, as well as in basic mathematics and physics. Still, *Telephony* magazine was moved to say in 1982 that while the Laboratories undoubtedly will continue to do first class research, "the benefits of that research may not reverberate through society as they have in the past."

And FCC economist David Chessler made the gloomy prophecy in the same year that "The loss of pure research due to reduction in Bell Operating Company payments to Bell Telephone Laboratories will eventually lower the growth rate throughout the American economy and make all U.S. electronics firms less competitive."

What is known for certain now is the primary mission of the Laboratories has been changed, and it must reluctantly give up its designation as a national resource. As Ian Ross, president of Laboratories put it in 1984, "Bell Laboratories has lost its end-to-end mission."

The boy steps up to the plate. Left foot and hip toward the pitcher, bat cocked over his right shoulder, determined and prepared with the intensity of a twelve-year-old Ted Williams.

The pitcher winds up and lets go. The ball zings toward home, accompanied by a slight beeping sound. The slugger calibrates height, distance, and WHAM, it's a hit!

A typical summer scene, played so often—except for one

detail. The batter was blind. He was able to judge the position of the ball by the little beeping sound coming from it courtesy of a Bell Laboratories-developed implant. The beeping baseball may have been a diversion for the scientists from more important work, but for that boy, it was the Laboratories' greatest achievement.

How many more such miracles will come from Ma Bell's House of Wonders? And how many miracles will be delayed or never achieved?

5

The Factory

> "... high-class service and low-class equipment do not coordinate."
>
> —THEODORE N. VAIL, 1915

Western Electric was an invisible giant.

The general public knew the Bell Laboratories. Its name popped up in newspaper articles with regularity—its new inventions, its scientists winning Nobel prizes, Washington asking it to coordinate vital projects.

But Western Electric? It was usually confused with Western Union or Westinghouse.

When Vail bought Western Electric in 1881, it was the largest producer of electrical goods in the United States. One hundred years later it had grown to be one of the ten largest American manufacturers, employing a quarter of a million people and selling $7.7 billion worth of its own products and another $2.3 billion in products made by other companies.

It was AT&T's factory. Western Electric took Bell Laboratories' research and development work and turned it into product. It mass-produced the most durable telephones on the market at half the cost of its closest competitor. It also turned out complex microelectronic chips. It was the world's first producer of the 64 kilobit chip and later the 256 kilobit chip.

But the general public, unless an individual was in the habit of looking at the bottom of his telephone, hardly knew

Western Electric existed, much less was aware of its immense size and scope.

The reason for this ignorance was twofold. One, AT&T owned 100 percent of Western Electric. So it was not listed separately on any securities exchange and its earnings and balance sheets were integrated with those of the rest of the Bell System.

Two, Western Electric didn't sell products directly to the public. It sold—with one exception—only to AT&T.

Of course, the Bell System was a gigantic customer, allowing Western Electric to have enormous production runs. Consequently unit prices were low. Also, because it had little competition, Western Electric didn't have to worry about cutting corners and could concentrate on quality production. Every Western Electric item was built to last virtually forever. Stories abound of vacuum tubes and relays in continuous service for more than sixty years. Electronic hobbyists put Western equipment high on their wish lists.

Having AT&T as its market was a mixed blessing. Western Electric prospered when the local Bell operating companies were growing and modernizing. On the flip side, a recession would bring a reduction in the local Bell's growth and Western Electric's business fell off dramatically. Unlike other companies in similar situations, Western Electric couldn't go looking for new customers to make up for the loss. Various government regulations precluded that.

When a local company's growth rate dropped from 4 percent to 2 percent, it was hardly a disaster for it. However, that 2 percent reduction represented a 50 percent drop in new equipment and cable purchases from Western Electric. The only saving grace was that even if a local Bell company had no growth at all, substantial facilities were still required to take care of telephone movement. In addition there were modernization and plant replacement purchases.

This volatility in sales led to volatility in employment at Western Electric. Layoffs were common.

The other Western Electric customer was also a large one. It was Uncle Sam who came knocking in times of national emergency.

The course of World War II might have been far different

without the research of the Bell Laboratories and the production of Western Electric. During that war both Western and the Laboratories worked almost exclusively for the military, leaving ordinary telephone business to get along as best it could.

There was no time for home telephones with Western Electric producing more than 50 percent of U.S. radar (about 57,000 sets of seventy different types) along with 30 percent of all its military communications equipment. Western was also producing the M-9 gun director, which played such a vital role in the war. Later Bell Laboratories developed and Western Electric produced the first electronically guided missile, the BAT, a homing bomb that was put to effective use against enemy shipping.

The end of World War II saw the beginning of the Cold War. President Harry S Truman appealed to AT&T to take over the Sandia Corporation, a company which had been directing the design, manufacture, and storage of nuclear weapons.

The Bell System was initially reluctant to take on this assignment since it was so far outside the communications field. But government pressure prevailed. Today Sandia, under overall AT&T direction and employing some seven thousand people, is still handling this work.

AT&T's involvement with the nation's defense did not stop with Sandia. Western Electric engineered and installed the DEW (Distant Early Warning) line, BMEWS (Ballistic Missile Early Warning System), SAGE (Semi-Automatic Ground Environment System), the White Alice over-the-horizon communications system, and the Nike family of anti-aircraft missiles.

The government knew that AT&T was one contractor that delivered. When national panic set in after the Russians launched Sputnik, the world's first artificial satellite, in 1957, Washington again turned to the Bell System. The United States was behind, national prestige was at stake, and the country had to catch up. Bell Laboratories and Western Electric got the first communications satellite up in 1960. This was the passive reflector Echo balloon. It was followed in 1962 by the active Telstar, which contained transmitting and receiving equipment.

Then President John F. Kennedy had a dream—a man, an American man, on the moon. And not in some distant century or even some faraway decade. He wanted fast action. He wanted it done in ten years. The U.S. would not only play catch-up with the Soviet Union, it would surpass it.

The undertaking, christened the Apollo project, was to involve massive planning. In 1962, the government asked Bell to direct systems planning for the project. With that end in mind the Bellcomm Corporation was formed, drawing largely from Western Electric and other Bell System employee pools. The rest was history.

So it was that on July 20, 1969, six months ahead of Kennedy's deadline, Neil Armstrong took that first small step for man. The Bell System could share the nation's pride in that accomplishment.

There was a tragic footnote to the Apollo project, one that might have been avoided if a Bellcomm recommendation had been heeded.

It had to do with using a pure oxygen environment in the Apollo capsule. Bellcomm scientists had counseled against it. They were worried about fire dangers.

Astronauts Gus Grissom, Ed White, and Roger Chaffee died when an Apollo capsule caught fire during ground tests on January 27, 1967.

Sometimes big is the only way to get things done.

There is a national mentality—no doubt brought on by the excesses of the Jay Goulds, the Cornelius Vanderbilts, and the Standard Oils of the last century—that big is bad.

As the song says, it ain't necessarily so. Some endeavors are not possible on a basement workshop scale. It took AT&T with its Bell Laboratories research and Western Electric manufacturing capabilities to design, build, and launch the first communications satellites. Telstar alone involved an investment of $60 million—in 1962 dollars.

While they are important to worldwide communications, satellites are not the be all and the end all. They have inherent limitations. Stationary satellites orbit at a height of around 22,300 miles. It takes noticeable time for signals to travel up and be returned. Trying to carry on a conversation, especially when more than one satellite connection is involved, often leads to people interrupting each other during the silent periods when the signal is en route.

So the Bell System poured money, lots of money, into an alternative—submarine cable. Again the system was geared up. By 1956 it had designed, manufactured, and installed the first transatlantic telephone cable. One of the most challenging aspects of the project was the more than 100 amplifiers that needed to be integrated with the cable on the ocean's floor. These amplifiers had to be built to last. Having to make frequent service calls to replace them at the bottom of the Atlantic was obviously not desirable.

This monumental undertaking cost some $50 million and was completely successful. The annual growth in the rate of transatlantic calling promptly jumped 25 percent.

This first cable only had thirty-six voice channels and was soon working at capacity. More cables were laid across the Atlantic and then the Pacific.

Meanwhile, an ingenious development called TASI (Time Assignment Speech Interpolation) was introduced in 1959. This multimillion dollar invention took advantage of gaps in conversation and even pauses between syllables. Other conversations got wedged into those gaps. With TASI, transoceanic cables could carry 50 percent more calls.

As big as these projects were in terms of cost, brain and manpower, they were dwarfed by another that consumed the attention of Bell Laboratories and Western Electric for two decades beginning in 1955.

Electronic switching.

Local and long distance dialing connections had been made using electromechanical switching systems that were expensive, slow, and took up a lot of physical space and electrical power.

Electronic switching was to change all that. And the key to its development was the concept of "stored program control." With this, an electronic memory, which could be easily altered, held the instructions for interconnecting calls rather than the masses of wires and relays used in older systems. In other words, the brains of the new machine was an electronic digital computer.

The project would have been much easier if it had been merely a question of developing and putting the new system into place. But it also had to work effectively with thousands of existing mechanical switching machines, some of them more than forty years old.

The first viable electronic central office switching system went into service in 1965. It took ten years of research, development—some failures—and more than $500 million. Without electronic switching, today's service demands could never have been met.

Still another large project was started in the early 1970s. This was the #4E four-wire electronic switching system, a time division switcher used to route long distance calls. In a four-wire system the transmitting and receiving paths are kept separate in the machine and not combined as they are in the more common two-wire switcher.

The 4E has no moving parts. All connections are made using solid state electronic components. Development cost? $400 million. Each machine's installed cost? $10 million. But each one can handle more than a half million calls per hour on more than 100,000 connected trunks. The future had arrived.

These are just some examples of what being big can accomplish. Some projects have immediate payoffs. After World War II, a plastic sheathing for cables was developed to replace the expensive lead covering used previously. This saved enough money in the Bell System to pay for the Laboratories' entire research budget for five years.

But other projects took years to pay off. The former Bell System could afford to carry the expense and go on to new projects. Big could be and was better at Bell. But therein lay the seeds of its destruction.

Vertical integration. One company handling the research, development, manufacture, engineering, installation, operation and maintenance—all the activities necessary to make a single phone call. Keeping these functions under one head gave ammunition to Bell's enemies. Western Electric's role, in particular, was to be a major factor in the demand for divestiture. Those gunning for Bell shouted "foul" over the one-supplier, one-customer relationship between it and the local

Bell operating companies. "Unfair advantage" and all that, they yelled because the local Bells were "required" to buy from Western Electric, thereby eliminating the possibility of any competition.

Those frothing at the mouth for Bell's blood could make that claim all they wanted—and they did—and they could also be dead wrong—which they were.

The truth was the locals were expected to buy much of their equipment from Western, but were not required to do so. Many outsiders being told this pooh-poohed the distinction as a bit of semantic smoke. They contended that any operating company purchaser who bought elsewhere was looking forward to one fat reprimand, or worse.

That was just not the case.

Western Electric had to tout and sell its products as did any number of other suppliers of similar goods. The AT&T staff sometimes issued cautionary advice against a non-Western Electric item, known as a general trade product. However, that wasn't to give Western some kind of extra advantage. Rather it was because the staff had not determined that the other product was good enough to get the Bell System seal of approval or that there would be limited maintenance support for these outside items.

Both authors were involved in many purchasing decisions for Pennsylvania Bell and know first-hand that Western Electric lost out on many occasions. Other companies got contracts for central office switching systems. Collins, Raytheon, and Nippon Electric sold microwave radio systems to Bell, and a whole host of other suppliers sold Bell test equipment and accessories. In each case, Western Electric representatives had touted the advantages of their products. In these cases, they didn't get the contracts. It happened.

Bell of Pennsylvania purchased almost none of its digital microwave radio equipment from Western for the same reason any other company would have passed it over. Western's competitors had better quality and lower-priced products. Period. Towards the end of the seventies and the beginning of the eighties, the same was true for digital switching equipment. Bell of Pennsylvania bought most of that from Northern Telecom.

No reprimands, harsh or otherwise, were ever handed down, nor were there even any suggestions from AT&T that this equipment should be bought from Western Electric.

On the other hand, it was true that local Bell companies did buy most of their equipment from Western. But for the same reasons they bought outside—on those items Western Electric had unbeatable quality and price. Western Electric supplied, almost exclusively, copper cable and wire, telephone instruments, multiplexing equipment, and most central office switching gear.

In 1972, the breakdown between Western and general trade products was 80 percent/20 percent. The Bell System spent approximately $1.2 billion that year on outside products. This increased gradually in subsequent years.

Top management was well aware that local Bell company purchasing decisions were being closely watched.

John deButts said in 1972, "I must emphasize how important it is that the Bell Companies' judgments as to what to buy and from whom be able to withstand the most searching public scrutiny with respect to their objectivity."

Keeping completely objective when it came to purchases was emphasized. Western Electric got the business when Western was the best. A further indication is that since January, 1984, when corporate ties between AT&T and the local Bells were slashed, the locals have continued to buy heavily from Western.

An interesting, if somewhat distressing, footnote to divestiture shouldn't be overlooked. When Bell was intact, Western Electric manufactured replacement and modernization parts for systems that had been installed up to fifty years earlier. As long as a system remained in service in the Bell System, it was essential it work with newer equipment which was coming on line. Keeping obsolete equipment working was not profitable for Western. It did so because that was part of the obligation of a vertically integrated supplier.

Western is no longer a vertically integrated supplier for the Baby Bells, so it's hardly reasonable to expect the same level of support to continue after divestiture. General Motors, after all, doesn't make replacement fenders for 1935 Chevys just because there are still a few out there. This will force the scrapping of some older systems, and this will translate into higher consumer costs. Just one more item the American public will have to foot the bill for because of divestiture.

Naturally, predivestiture, Western had the replacement parts field pretty much to itself—no other company was interested—along with the large market for additions to central office equipment originally made by Western. In other cases, such as telephone instruments and cable, Western had a cost advantage over its competitors because of its mass purchases of raw materials and enormous production runs. But even more than that, Western's winning card was the close association with Bell Laboratories. Together they were able to produce quality, and quality is hard to beat.

Because it had these markets to itself, there was the potential for Western to play price games: to inflate the price of items where it had no competition so it could lower the price where it had.

ITT filed an antitrust suit charging just that. AT&T eventually settled out of court, agreeing to buy more ITT equipment. But Washington decided that where there was smoke and lawsuits, there must be fire. So it launched numerous investigations, spending millions, searching for evidence that Western was involved in unfair trade practices. Apparently it was inconceivable to the bureaucratic minds at the FCC that the company could actually produce superior products at low costs honestly.

Lo and behold, the investigations failed to turn up any evidence that Western had, in fact, played such games. Nothing. Despite all those millions of dollars spent, repeated analyses showed that Western Electric prices were below those of its competitors simply because of production efficiencies and volume.

The strengths of the old Bell System were to become its Achilles heel. Critics could not believe that Western Electric products could cost so little and be so good without some sort of chicanery. Bell's top managers, brought up and instilled with corporate integrity, were not prepared for down-dirty streetfighting or the disaster that was on the horizon. Who in their right mind would endanger a national treasure such as Western Electric, particularly in combination with the Bell Laboratories? Who indeed?

John deButts, the AT&T chairman knew there was trouble in 1977. He predicted that "the [Federal Communications] Commission's policies ... will force an alteration of our research and development priorities ... to give more attention to relatively superficial product differentiation and—reluctantly—somewhat less fundamental systemic improvements of advantage to our entire customer body."

But in retrospect that prediction seems almost laid back and nonchalant. A breeze had started blowing and Ma Bell hardly noticed it.

6

The Beginning of the Storm

The winds began picking up.

Theodore Vail probably assumed the 1913 Kingsbury Commitment settled the antitrust bugaboo.

By letting independent companies plug into the Bell System and by agreeing not to buy up any more independents without federal approval, it was believed by Bell managers that Washington would stop gunning for AT&T under the Sherman Act.

It seemed pretty clear, when reasonable, informed people thought it through, that the telephone system was a public utility, a natural monopoly much like a gas or electric company. Why? Because a natural monopoly exists when the economies of scale dictate that one company alone can provide the customer with a service or product at a lower unit cost than two or more competing companies could. There are enormous underlying structural costs in the telephone transmission and switching network—the land, buildings, switching systems, power plants, right-of-way, conduit, cable, microwave towers, etc. Two companies competing means a duplication of all this—and two sets of facilities aren't needed to provide service.

Efficiency, then, is another key element in natural monopolies. Is it more efficient—and therefore more beneficial to the public—to eliminate competition and have only one company in the market?

Economists through the years have accepted telephone service as a natural monopoly. In citing the greater efficiencies of these monopolies, Harold Koonitz of the University of California at Los Angeles and Richard W. Gable of the University of Southern California wrote, "If competitive telephone, gas, water, electricity, or street-railway lines were allowed to build multiple lines down the city streets or across property, the result would be a confusing and inefficient mass of duplicating facilities. These might not only hamper use of the streets but impair the quality of service."

There's no getting around it, the telephone system is a natural monopoly. Once that's accepted, what British economist John Stuart Mill had to say 140 years ago about introducing competition into such a monopoly has to be considered. It was his widely accepted contention that when competition is introduced into a public utility operation the consumer must pay for the sum of the costs of all the competitors. In other words, plain and simple, the consumer is being gouged, taken, and ripped-off. And that is wrong.

Because of its wastefulness, Mill argued, competition in a public utility should be prohibited for the public good. However, the public utility monopoly must be carefully and properly regulated to avoid the user being overcharged with the regulators keeping in mind that the rate structure should be based on the principle of the "greatest good to the greatest number." Those who can afford to pay more should pay more so that the lower economic classes may also be served.

With the Kingsbury Commitment, it was believed that Washington had accepted AT&T's status as a monopoly, and any future complaints the government had with its operation would be resolved through regulation.

But the commitment was only a lull—granted a long one—in the antitrust storm. Washington doesn't stay the same. New administrations bring new personalities and new attitudes, and few so dramatic as Franklin Delano Roosevelt's New Deal.

Up until then, federal regulation of the telephone industry had fallen under the umbrella of the Interstate Commerce Commission, which was really more interested in railroads than anything else. In 1934, Congress created the Federal Communications Commission. One of its primary functions was to oversee the interstate traffic of the telephone industry.

The first thing the chairman of its telephone division, Paul Atlee Walker, a reform-minded Quaker, did was start an investigation of all telephone companies that was soon narrowed down to only Bell.

AT&T president Gifford declared, "We welcome the investigation; there are no skeletons in our closet."

The proceedings were long, exhaustive, and Walker was clearly biased against AT&T. When he finally issued his "Proposed Report"—largely his work and not that of the other FCC commissioners—it was an out-and-out attack on the company. He was especially upset over Western Electric's role, calling for competitive bidding in the purchase of equipment.

The final report, which came from the full commission and not just the fervent Walker, was considerably toned down—and virtually overlooked. The country was far more concerned with the frightening events taking place in Europe in 1939 than it was in who should manufacture telephones.

After the war, Holmes Baldridge, an attorney who had worked with Walker on the investigation, joined the Justice Department's antitrust division. Baldridge agreed with his former boss that AT&T's relationship with Western Electric was bad for the country. He pushed for and convinced Attorney General Tom Clark to file an antitrust suit in 1949, calling for the separation of Western Electric and Bell Laboratories from the rest of the Bell System.

Of course, some people found the suit illogical, coming at the same time the government was begging the Bell System to take on the management of the U.S. nuclear weapons facility in Sandia, N.M.

As AT&T President Leroy A. Wilson pointed out in 1949, "What the government asks in this lawsuit is that the courts break up and dissolve the very organizational unity and size this vital security job requires. We are concerned by the fact that this antitrust suit seeks to terminate the very same Western Electric/Bell System relationship which gives our organization the unique qualifications to which you refer."

Over the next seven years the suit ran a rocky and controversial course. Finally in 1956, an agreement in the form of a consent decree ended the suit.

At first glance, the agreement seemed to provide a basis for stability in the industry. Bell retained Western Electric and Bell Laboratories. In return, the company agreed not to branch out into other fields. It had to confine its activities to common carrier communications. Further, Western Electric would have only one market—the Bell System. And lastly, license and technical information developed in the Bell Laboratories would become available to any who applied for it.

These restrictions didn't seem too severe and most Bell System managers were pleased with the outcome.

"We kept Western Electric," Frederick R. Kappel, AT&T's president at the time, said, "but at some cost patent-wise. The decree ... generally makes legal an integrated Bell System."

The unity of Bell had been preserved, although it took a while for the controversy surrounding the decree to subside. It was charged, for one thing, that Attorney General Brownell and the Eisenhower administration had caved in to AT&T pressure. A Congressional investigation was launched, but it did not lead to a reversal of the settlement.

So, all looked rosy for AT&T. Who then could have guessed that this victory was actually the first inexorable step towards defeat? It would have taken a crystal ball to have divined the land mines put in place by the consent decree. One was limiting the Bell System to common carrier activities. That seemed straightforward enough in 1956. However, with the onslaught of the computer revolution the difference between data processing (a no-no for Bell under the decree) and data transmission (allowable) became less and less distinct. This created problems for both Bell and its regulators.

In fact, the FCC would eventually launch three inquiries— Computer Inquiry I, II, and III—to establish rules for separating regulated communications from unregulated data processing, as the boundaries between them became fuzzier with the introduction of new technology. And even after all that, few were satisfied with the results.

The problems began around 1960 when the concept of

computer time-sharing was invented. An early application involved the composing of a message at one computer terminal and having it printed at another.

Whoa! cried Western Union to the FCC. The computer companies offering these services were illegally engaging in common carrier communications. Western Union countered with a similar service.

Just one second there! screamed the computer data-processing firms. Western Union was going beyond its charter as a regulated common carrier and was competing unfairly by entering the unregulated information-processing business.

Who was right? The FCC didn't know, so in 1966 it set out to unravel the situation by establishing some rules for separating regulated communications and unregulated data processing. This proceeding was later called Computer Inquiry I and didn't conclude until 1971. One rule it came up with was any service that fell on the borderline between communications and processing of data would be considered a "hybrid" service. The FCC would then decide, case by case, whether or not the service should be regulated.

It wasn't long before the commission was inundated by hybrid cases, many of which turned into long and bitter debates between the protagonists on both sides.

By 1976, the FCC decided it was time for some better definitions and rules. And so was born Computer Inquiry II. Out of this second inquiry, completed in 1980, came the notion of two classes of services—basic and enhanced.

Basic services were defined as the simple transmission of information without modification. These could be offered by existing common carrier companies such as AT&T, GTE, and Western Union.

Enhanced services were defined as those that acted on the content or format of transmitted information, and they encompassed videotext, protocol conversion, information storage, packet switching, and other modifications of transmitted data. These could only be offered by companies outside the regulatory arena or fully separated subsidiaries of AT&T. It is interesting to note—and perhaps to wonder why—only AT&T and not other common carriers had the separate subsidiary restriction. This ruling forced a costly organizational restructuring on AT&T in the midst of its other problems.

Computer II also required that all customer premises equipment, including ordinary residence telephones, must be sold through separate subsidiaries rather than continuing to be included as part of tariffed services. This was to get around any bickering over whether a particular type of equipment was a communications terminal or a data-processing device.

So the problem seemed to be settled at last. Except. Except the separate subsidiary solution created its own set of problems. Certain advanced services such as centralized answering and recording couldn't be introduced. The physical separation of the equipment providing these services from the equipment providing basic telephone service was impossible. AT&T had to scrap the combined equipment it had already installed in several locations at the cost of several million dollars. While the separate subsidiary requirement might have been well-intentioned, it ended up costing the public advanced services and a lot of money for AT&T. AT&T has estimated the separate subsidiary requirements alone were costing it more than $1 billion annually.

Naturally, this led to Computer Inquiry III in 1984. One of its results was a 1986 FCC decision essentially eliminating the subsidiary requirement and substituting accounting rules to prevent subsidization of nonregulated services.

All of this description of the inquiries merely shows that what seemed clear-cut in 1956, wasn't.

Another land mine proved to be the decree's preventing Bell from following the natural paths opened up by its own technology into other lines of business such as computers. At the same time, the decree armed future Bell competitors with Bell's own technical information.

Still, what was important in 1956 was that the consent decree had reaffirmed the principle that the network over which calls are transmitted is a natural monopoly. Or at least that's what Bell managers thought. But it was only three years later in 1959 that the FCC made the first significant break in the natural monopoly principle with its Above 890 ruling. This allowed private companies to use available portions of the radio frequency spectrum above 890 megahertz for their own communications.

As it turned out, the ruling itself had little impact. Most companies found that using the Bell System for their internal communications was cheaper and more reliable than constructing and maintaining their own networks.

But the FCC didn't stop with Above 890. In June, 1968, came its Carterfone decision. The Carter Electronics Corporation of Texas manufactured a device that could connect private two-way radio systems with the telephone system. In other words, it wanted to introduce a "foreign" piece of equipment into the network. This was something AT&T had always fought against, arguing that the integrity of the entire system could be damaged by plugging in inferior equipment. The Bell System's tariffs had prohibited such interconnection.

Carterfone appealed to the FCC, and the FCC told AT&T to revise its tariffs.

A year later, the FCC went further. In August, 1969, it authorized a common carrier, Microwave Communications Incorporated (MCI) to construct point-to-point microwave radio systems and to sell private line services to individual business users. Initially, MCI said it was only interested in providing service between Chicago and St. Louis. MCI's chairman, William McGowan, specifically stated that his company had no intention of going beyond customized point-to-point service. MCI had no desire to compete with Bell System's long distance network.

That's what William McGowan said. So the FCC approved MCI's application, apparently believing that the company only wished to provide innovative services not available from AT&T. But what McGowan said and what McGowan did were two very different things. Within a shamefully short time, and much to the surprise and consternation of the FCC, MCI without any specific authorization starting offering long distance service in direct competition with AT&T.

The FCC was a bit upset and ruled that MCI's long distance service, called Execunet, was illegal. However, the federal appeals court begged to differ, and overruled the decision in July, 1977. The local operating companies were then required to give MCI connecting links for its long distance calls.

Both Carterfone and MCI, it was to turn out, were the first steps toward opening the telephone network to unrestricted competition. Why did the FCC want to do that?

The commission was apparently working with five assumptions that were largely fallacious and did not take into account technological and economic realities. The FCC, after all, was, for the most part, a group of lawyer-bureaucrats, not engineers, not scientists, not economists. The assumptions, it would seem, were:

1) To regulate interstate communications, the FCC must regulate all communications.

2) Competition is preferable to regulated monopoly.

3) Big is inherently bad.

4) Charges for all services should be based on the actual cost of providing those services.

5) Customers should be encouraged if not required to own and maintain their equipment.

At a fast read, some of these assumptions might appear reasonable. However, none of them stand up to informed scrutiny. And further, they led to the destruction of the Bell System at an insanely high cost to the nation—and not just in dollars and cents.

By making that first assumption, the FCC waded into deep and murky constitutional waters. The Communications Act of 1934 chartered the FCC to oversee and regulate interstate communications. Communications within states (and in a few instances in cities and counties) were regulated by local commissions. This arrangement was to be expected under the constitutional separation of powers between individual states and the federal government.

But the two hundred years since the Constitution was ratified have seen a gradual erosion of the states' authority with a concomitant increase of power in the federal government. Similarly, over a much shorter period of time, the FCC has been usurping the authority of various state regulatory commissions.

One justification for more federal control of telecommunications was the occasional difference in the cost of a call depending on whether it crossed a state line. For instance, at one time it cost more to call from Camden, New Jersey, to Newark, New Jersey, than it did from Camden to New York City, even though New York was farther away than Newark. The intrastate rate was set by the New Jersey Public Utilities Commission, the interstate by the FCC.

If all rate setting were to be done by the FCC, it was argued, such disparities would be eliminated. However that argument loses some credibility when it's realized that these and other disparities were well on the way to being eliminated through public pressure and without FCC intervention. One has to wonder whether the FCC began nibbling away at the states' powers out of a sense of fair play and concern for the public or whether it was just another case of bureaucratic empire building.

The FCC didn't send out an announcement one day that it was taking over from the states. It was more like ice-age glaciers inching down the countryside. Initially the FCC's control was largely through setting of depreciation rates. These rates affected net corporate earnings which were inevitably translated into the rates allowed by the state public utilities commissions. But the ante was upped late in 1973 when the FCC notified AT&T that state tariffs must be filed with the FCC. Following hard and fast on that, in January, 1974, the FCC announced that state commissions could take no actions on interconnections that were not in accord with federal policy. This took 25 percent of local companies' investment out of state control.

States did not seriously challenge these rulings or those that followed even though they weakened the authority of their own agencies. Perhaps they didn't understand what was happening. Only recently has there been an outcry from state capitals, then only after the FCC slapped access charges onto local telephone bills. The local bills had been the sacrosanct domain of the state.

Ironically, the access charge was forced onto FCC by divestiture. The local operating companies no longer got subsidies for local services from AT&T's profitable Long Lines. To offset this loss, local rates had to be increased by something called "customer access line charges." Thus, by forcing competition into long distance service, the FCC created a problem that it could only solve by, in effect, taking complete federal control over telephone rates. AT&T, after all, was out there competing for long distance customers with MCI, Sprint, and the others as the FCC had deemed desirable. It couldn't be expected to subsidize local service any longer.

Which leads to Assumption Number Two. "Competition is preferable to regulation."

As a people, Americans are competitive. We tend to equate competition with freedom. It's our inalienable right to have an equal chance at the gold ring. Regulation, it then follows, is linked with oppression. In our open society this reasoning is often true. However, most people would agree that it is not desirable to have two electric companies serving the same community. The duplication of power plants and distribution facilities not only wastes money, it's ugly.

So rather than have unrestricted competition, electric companies are given exclusive franchises for their services and their rates are regulated by government agencies.

Alfred E. Kahn of Cornell University and former chairman of the New York State Public Service Commission has defined a natural monopoly as existing if "the service is such that the consumer can be served at least cost or greatest net cost benefit by a single firm."

And that's where the controversy lies. Is telecommunications a natural monopoly? Will the public be best served if it is accepted as such?

Without going into great detail at this point, it can be said that some parts of the telephone business—exchange distribution, for instance—are clearly natural monopolies. Other parts—equipment manufacture—are not, and should be subject to competition. But in areas such as network services and terminal equipment, the lines get fuzzier and debate heats up.

Up until the late 1960s, the FCC treated the industry as a monopoly, subject to regulation. But then, without announcing a change in policy, it began substituting competition for regulation through its decisions. These actions, of fundamental importance to our nation's communications, were never explicitly debated and got very little public attention. But more incredibly, these decisions were made with little or no input from qualified economists or engineers. The lawyers and the bureaucrats of the Justice Department and the FCC made them without even examining the possibility that a regulated monopoly in communications could serve society better than unrestricted competition. This wouldn't be much different than throwing all the scientists out of NASA and letting a bunch of attorneys try to launch the next space probe.

What's more, competition in the long distance field was only possible when funded by the biggest steal in United States history. MCI, Sprint, and the others could not have competed with AT&T without help. It was economically impossible. The actual cost of providing a long distance connection through MCI *et al.*, is from two to four times higher than a similar call on AT&T's network. But the MCIs were able to undersell Bell with the connivance of the FCC. For the commission required AT&T to pay the local operating companies and the independents $400 million for every $1 billion it earned in long distance revenue. What did the FCC require MCI and the others to pay the local companies? Next to nothing in comparison.

But why did MCI and Sprint deserve this boodle? Because competition—at whatever cost, it would seem—is better than regulation, naturally. Though Gene Kimmelman, director of the Consumer Federation of America, disagrees. "Regulation has gotten a bum rap," he said in 1987. "It has provided cheap service and innovation." But go tell that to the FCC.

Assumption Number Three. Big is bad. No one can argue that big can be bad. The muckrakers such as Ida Tarbell documented that. Out of the excesses of the oil and railroad barons came the Sherman Antitrust Act of 1890 and the Clayton Act of 1914.

The antitrust division of the Department of Justice got the job of seeking out and prosecuting suspected violators of these laws. Over the years, its tendency has been to go after the largest and most visible targets.

There are two sides to every coin, something the Department of Justice at times has apparently forgotten. There is no doubt that a big corporation has more opportunity and capability to be bad than a smaller one. But it doesn't necessarily mean it will be. Also a big corporation has far more capability to be good. Being big and good can translate into greater efficiencies and lower prices, into more basic research. It can also extend to outside activities such as support for social welfare, education, or the arts.

Why label an enterprise "bad" because it's big and therefore has the potential to be bad? If that were so then anyone owning a gun could be called a criminal because someday, maybe, that person could use the weapon to commit a crime.

During the 1970s, the FCC seemed to accept outright the "big is bad" principle and applied it to Bell. Not only would the agency stop growth at Bell, it was determined to reduce its size. The concept of economy of scale apparently was never part of its thinking.

To prove that Bell was indeed bad, the FCC spent millions of taxpayer dollars on several investigations. Nothing significant was ever unearthed. One of the authors (Constantine Raymond Kraus), after retiring from Bell and setting up a consulting firm, was hired by the government to conduct an investigation. Specifically, he was asked to determine whether Western Electric prices were being set properly and whether the company was adhering to approved depreciation accounting practices.

He searched, but couldn't find anything of great import. In fact, the most significant aspect of the investigation was that the government never publicized the results.

Touche Ross, a highly respected consulting firm, was later hired by the FCC to do a complete review of Western Electric. In January, 1974, after a thorough investigation, Western Electric got a clean bill of health. Moreover, Touche Ross concluded that Western produced equipment efficiently and at lower costs than it could if it were not part of the Bell System. It also concluded that the relationship between Bell Laboratories and Western Electric was ideally suited for large-scale technical innovation at the lowest cost to customers.

And lastly, Touche Ross stated that Western Electric and Bell Laboratories should not be separated from the Bell System.

This report was directly opposed to the direction the FCC

had been taking. In the case of Bell, big wasn't bad, Touche Ross was saying. On the contrary, it was good.

The FCC never published the findings of Touche Ross. If it had, how could it have justified what was to follow?

Assumption Number Four. The charge for a service should be based on what it costs to provide that service.

Bell had based its pricing philosophy on John Stuart Mill's concept of "the greatest good to the greatest number." The cost of the service did not matter as much as the value of the service.

This meant that a telephone call of given distance, duration, and time of day had the same charge as another call of similar distance, duration, and time of day even if one call actually cost more for the company to provide because of the way it was routed. Indeed, one call might represent a large profit to the company and the other a loss. Similarly, the monthly charge for a particular grade of service in a community was the same whether the customer lived next door to the central office or five miles away. If the value of service to the customer was the same—each one was getting the same service as far as they were concerned—then the charge was the same whether or not there was a difference in cost to the company.

This pricing philosophy led to universal telephone service in the United States, unheard of anywhere else in the world, although Canada came close. Historically, the FCC had condoned this pricing system. But its rulings in the 1970s clearly indicated it was abandoning the value of service concept for cost-based pricing. The Bell System read the writing on the wall and frequently argued that cost-basing of rates would undermine its system of internal subsidies that made telephone service affordable for everyone.

Bell analysts predicted that cost-based pricing would lead to a 50 percent reduction in long distance rates. Great for businesses and more affluent resident users. But it would also increase local rates by 100 to 500 percent, socking it to the poor who would be forced to discontinue their service.

Bell argued but the FCC wasn't listening.

The FCC push for cost-based pricing meant that individual

segments of AT&T's operations would have to be analyzed more closely. This meant that Bell's ranks of these analysts, known as cost study engineers, swelled from less than 100 in 1972 to nearly ten times that a decade later. The additional cost in salaries alone—forget about office space and paper clips—approached $50 million a year.

The emphasis on cost-basing led to the development of something called "functional accounting," a complex computerized procedure that was supposed to relate every cost the Bell System incurred, no matter how small, to the activity causing it.

It cost almost a billion dollars to develop this functional accounting system. It cost even more to operate it. But not a penny of this in any way improved customer service. All it did was increase accounting costs in an effort to appease the irrational demands of the FCC.

In the end there was no pleasing the FCC. AT&T thought it could meet the commission's cost-basing demand, at least part way, by charging more per mile for transmission over lightly used routes than over denser ones. The FCC shot down this Hi/Lo tariff filing in January, 1974.

Over the years the FCC seldom gave AT&T any guidance on pricing procedures. One tariff filed by AT&T, for TELPAK service, languished in the bowels of the FCC for twenty years without being approved or disapproved.

Assumption Five: Customers should own and maintain their telephone equipment.

Let's say a person, Mr. B., wants toast and blueberry jelly for breakfast. (Bear with us, this does have something to do with telephones.) So Mr. B. goes to Sears, pays $24.95, and takes his two-slice toaster home. He's paid his electric bill, so he plugs the toaster into the outlet, pops in his bread and within minutes, voila! Toast.

In the past, if Mr. B. wanted to call Aunt Gladys in Peoria, he had to contact the telephone company, which gave him end-to-end service. It hooked him into the system and it "lent" him a telephone. Then Mr. B. could call his beloved aunt.

To the FCC, this wasn't right. Why could Mr. B. own his toaster but couldn't own his phone? Why shouldn't Mr. B. be allowed to buy a telephone and plug into the telephone system and get the "juice" to run his phone just as he plugged into the electric lines?

Any engineer could have explained why not. It's really very simple.

Say that toaster from Sears, unlikely and as improbable as this may be, was a faulty toaster. And Mr. B. unknowingly plugs it in. Zappo. He blows a fuse or his circuit breaker shuts off, or, even worse, he injures himself or his property. That's the extent of it. His bad toaster shorting out Mr. B.'s electricity isn't going to affect the house down the block or the rest of the electric company's system. That's because Mr. B. receives electric energy but feeds nothing back into the distribution network. It's a one-way, incoming proposition.

That's not how the telephone system works.

Let's say Mr. B. covets the buffalo-shaped telephone on sale at Cheapo Charlie's. He buys it for $9.95 and brings it home. When he plugs it in, he's interconnecting with one large computer. And his little buffalo is a computer terminal. This is a two-way street. Mr. B. is receiving input from the system and also transmitting into it. When his crummy buffalo telephone gives up the ghost, that not only affects his being able to call out but also keeps poor Aunt Gladys from calling him, affecting her service.

Worse yet, equipment that is faulty or improperly operated can cause malfunctions and overloads in the central office switching equipment and that can affect everyone.

Despite all this, the FCC never could tell the difference between a toaster and a telephone.

Former FCC Chairman Mark Fowler once set forth two profound thoughts. One was that standards are unnecessary in the telephone industry. The marketplace, he said, would determine the standards. Of course, one may wonder how the uninformed lay customer is supposed to know what the standards should be. But if that thought makes little sense, consider the chairman's second one, that out of the chaos of competition would emerge the cheapest and best service for the customer. With any understanding of the telephone network, it is clear that the only thing to emerge from the chaos of competition will be chaos.

But it was based on such fallacious thinking that the FCC

decisions paving the way for competition were apparently made. The rulings were many, separate, and often quite narrow. Nonetheless, they brought on the complete restructuring of our telecommunications network.

Perhaps what happened would be easier to accept if there were any evidence of the use of legitimate analysis, master planning, or any kind of overall strategy.

But the evidence there is points to one disturbing conclusion.

The FCC was winging it.

7

The Tempest

While the FCC was merrily slipsliding down its twisted road to competition during the sixties and seventies, some of the longtime antitrust division lawyers at the Department of Justice were still smarting over Attorney General Brownell's consent decree of 1956.

There had been an unpleasant taste in their mouths, a taste of the department selling out to big business, that they had never quite gotten rid of.

They didn't believe the FCC could regulate AT&T any more. With an influx of young, liberal attorneys, hot off the radical campuses of the sixties, new life was breathed into the Department of Justice's old case against AT&T. Some may have seen the case as a chance to make headlines and boost careers. Others no doubt truly believed they were riding to the rescue of a public being victimized by a "predatory monopoly."

Investigations were again launched, after a good deal of intensive lobbying by MCI's president William McGowan, not exactly a disinterested observer. The Justice Department, for the third time, felt it had a solid case against the Bell System. It would charge that AT&T had tried to monopolize the long distance market by denying interconnection or making it hard for competitors to interconnect. That it had smothered equipment competition by forcing the local Bell companies to buy from Western Electric.

In some respects AT&T had been its own worst enemy. For instance, when the FCC decided that non-Western Electric customer premises equipment could be used as long as Bell provided protective interfaces, or couplers, Bell didn't rush to make these devices available.

There was a certain arrogance at AT&T, based partially on the belief that the federal government, when push came to shove, would not destroy a system that worked as well as Bell's.

The Department of Justice saw things differently and got its case together.

But this was 1974, and this was Watergate, and the Department of Justice was leaderless. (Had it not been, the department's lawyers might not have had the autonomy to carry on the investigations as they did.)

John Mitchell was out as attorney general. His successor, Elliot Richardson, had resigned in October, 1973, rather than fire Special Prosecutor Archibald Cox. And then for a while Nixon had too many other things on his mind to find a replacement for Richardson, so the department lay becalmed in the water. Then he named former Ohio Senator William Saxbe to the post.

Saxbe, it's been reported, wanted to clean up the image of Justice, which had been so tarnished during the Nixon years. What better way than don the white hat and haul some Big Bad Guy into court with a highly visible case. Saxbe decided to go ahead against AT&T.

There is some dispute as to whether Saxbe consulted Gerald Ford, who by then had replaced Nixon. What is known is that on Wednesday, November 20, 1974, while Ford was in Japan, the Department of Justice filed an antitrust suit against AT&T calling for a breakup of the gigantic company, and more specifically, for the separation of Western Electric and Bell Laboratories from the parent firm.

There is no such thing as a "speedy" trial in antitrust litigation. The federal government had gone after IBM, the country's largest computer company, six years earlier, and that case wasn't even close to the courtroom.

The size of the suit brought by Justice against AT&T was mind-boggling. It's believed that the 1974 filing was the largest such action on record and it didn't reach the courtroom for six and half years. It's estimated that seven billion—billion!—pages of material were examined of which one billion—billion!—were copied and entered as evidence.

AT&T created a complete facility in Orlando, Florida, just to collect and organize the paperwork. More than one thousand people were assigned to this job. Another three thousand AT&T employees worked full time on the defense, and thousands of others spent varying amounts of time, often during evenings and weekends, on the case.

It's impossible to say what all this activity actually cost, but the government and AT&T have admitted to direct charges of more than $400 million. That's direct, not total costs. A company with fifteen hundred employees was created solely to handle the splitting and transfer of stock in AT&T and the divested companies. An outfit named the American Banknote Company received more than five million dollars just to print the certificates.

Take one guess who in the end paid that bill. The taxpayers—which most Americans are—picked up the government's check and the telephone users—which most Americans are—AT&T's. But then again, who can complain? Think of all the social benefits that came out of this action—the salaries and fat fees for the army of lawyers, the thriving business for printers and paper manufacturers, the scads of extra tickets sold by the airlines who flew witnesses to and from hearings, to name a few.

Finally the years of preparation and legal skirmishes were over and it was time to enter the courtroom.

The case had been assigned to federal Judge Joseph Waddy. AT&T attorneys were hopeful. Judge Waddy appeared initially unimpressed by the government's case. But fate intervened, as if in some Greek tragedy, and Judge Waddy died of cancer. His cases were divided among other judges, and AT&T found itself facing Harold H. Greene, and its lawyers lost a bit of their hope.

Greene was new to the federal bench, but he had a strong, liberal background. The judge, whose family had fled the persecutions of Hitler's Germany, had worked under Bobby Kennedy in the Justice Department and had been instrumental in writing the Civil Rights Act of 1964 and the Voting Rights Act of 1965. He had been on the lowly District of Columbia municipal court for years before the Carter administration called him up to the federal bench.

When Greene got the AT&T case, he told lawyers from both sides he wanted to demonstrate that antitrust laws were enforceable, that business couldn't endrun them by pulling interminable legal ploys out of the huddle and keeping the cases tied up in court forever.

It was not an auspicious beginning for AT&T. Would Greene be showing that antitrust laws were enforceable if he decided for AT&T? Was he telegraphing from the start that AT&T didn't have a prayer?

During pretrial, both sides not only expended vast amounts of time and money assembling evidence, they looked for ways to settle out of court that would be acceptable to both AT&T and the Justice Department. Legislative solutions were considered. Several abortive attempts were made, some even initiated and strongly supported by AT&T, but they were all stalled in Congress. In early 1981, it looked as if one bill, S-898, actually had a chance for passage.

Both AT&T and Justice agreed that if Congress passed a suitable bill, it would be desirable to drop the antitrust suit. Judge Greene was petitioned to delay the trial to see what Congress would do with S-898. Greene refused. The best guess was he was angered by Justice's on-again, off-again approach to the case.

As it turned out, the bill was amended so many times that by the time it was finally passed by the Senate in October of 1981, it was no longer acceptable to AT&T. It didn't matter. S-898 never got out of the House.

Meanwhile, back at the White House.

The U.S. vs. AT&T, it should be remembered, was filed, went to trial, and was finally resolved under the administrations of two pro-business Republican presidents. Either man, as president and head of the executive branch under which the Justice Department falls, had the authority to stop the action. Future historians might scratch their heads over their failure to do so unless they keep in mind some extenuating circumstances. Both presidents, at the crucial times, were deeply occupied with other matters.

Late 1974, when the suit was filed, was a critical time for the office of the presidency and for the nation. As a result of the Watergate scandal, Richard Nixon became the first president in U.S. history to resign from office. One of Gerald Ford's first exercises of executive authority was to grant Nixon a full pardon. While Ford might have intended to put the Watergate affair behind and get the nation on to more pressing matters, he only managed to generate a storm of public protest and anger. This pardon may have well cost Ford election two years later.

Many Americans were infuriated by Ford's action, seeing it as interfering with due process of law. Once the AT&T suit was filed, Ford undoubtedly felt little inclined to incur additional wrath by squelching the case.

But what of Ronald Reagan? He was a man, who as candidate running for his first term, had used AT&T's good service and low rates as examples of what private enterprise could accomplish. He often contrasted the reductions in telephone rates over the years with rising cost of a first-class stamp to illustrate his theme that big government was harmful to the nation. And he'd add, "Of course, the government is suing the phone company."

Surely Reagan was against breaking up the Bell System.

Again, circumstances intervened. First, the case went to trial in January, 1981, within a week of Reagan's inauguration. The first months of any new administration are filled with confusion and high priorities. But there was more than usual transition chaos involved in this case. For one thing, both the newly appointed attorney general, William French Smith, and his deputy, Edward Schmults, had to disqualify themselves from participating in the case because they had been once closely involved with the Bell System.

It was not until late February that the next person in Justice's chain of command was appointed. He was Assistant Attorney General William F. Baxter.

Now here the plot thickens—and gets preposterously muddied. Baxter was a conservative, free-market advocate who had been known to call the U.S. Supreme Court "whacko," and who held that the Justice Department "penalized [big companies] because of their size." In 1977 he had said, "I think the telephone company is telling us the truth when it says that if more competition emerges in the specialized carrier areas in long lines, telephone rates are going to have to go up on the local loops at the expense of business use of long lines communications." This indicated he was aware of the adverse effects of competition.

But more importantly Baxter had also written that the AT&T case was "the one good thing the antitrust division has done in the last thirty years."

It's believed the Reagan administration was unaware of this quirk in Baxter's thinking.

To complicate matters even further, Baxter and the new Secretary of Defense, Caspar Weinberger, had a falling out caused by a misunderstanding.

Weinberger was one of at least three Reagan cabinet members—Malcolm Baldrige of the Commerce Department and John Block of Agriculture were two others—who strongly opposed the breakup of AT&T. Weinberger, not knowing that William French Smith had withdrawn from handling anything involved with the case, sent the Attorney General a strong letter urging that the suit be dropped. Along with it was a letter from the Joint Chiefs of Staff discussing military reasons why dismantling AT&T would be disastrous to the nation's defense. Some of the information in that letter was classified. Since Baxter had just been appointed, he had not yet received his security clearance.

Weinberger's letter, therefore, was put into a safe pending Baxter's clearance. When the Justice Department did not drop the suit, Weinberger thought his letter had been read and ignored. He testified before a closed hearing of the Senate Armed Services Committee about his call for dropping the suit and added that he had written the letter to the Justice Department.

As it happened the *Wall Street Journal* got hold of the secret testimony and ran a story on it. Baxter was angered and held a press conference in which he announced, "I do not intend to fold up my tent and go away because the Department of Defense expressed concern." He added that he thought the case had a "sound theoretical core, and I intend to litigate it to the eyeballs."

Ronald Reagan, who favored a management style of consensus, had disagreement in his ranks. Perhaps he favored dismissal but wanted agreement among his subordinates. Perhaps he feared dismissal at that point, with all the baggage the case carried with it, would look as if someone was being paid off and he didn't want his own Watergate before his chair in the Oval Office was even warm. Maybe ... maybe Reagan will explain when he gets around to writing his memoirs.

What is known is the case went forward.

By this time, John deButts was no longer chairman of AT&T. The more unobtrusive, soft-spoken Charles Brown had taken his place. He was to be the leader of the defense with Baxter leading the prosecution and Greene in the role of judge and jury. The phalanxes of courtroom lawyers were headed by George Saunders for AT&T and Gerry Connell for the government. And the cast would not be complete without mentioning the hundreds of witnesses who were called and cross-examined by both sides.

The case went to trial in January, 1981, with the government presenting its arguments first. The testimony was enough to fill a small library. At the heart of the government's case was the claim that AT&T had engaged in anticompetitive practices, specifically by stonewalling on interconnections with MCI and others, and trying to restrict connections of non-Western Electric equipment. And it threw in for good measure the contention that the FCC was incapable of regulating AT&T as the company was then structured.

AT&T argued that all its prices, services, and actions had been monitored and regulated by the FCC and the various state commissions and that AT&T had leaned over backwards to comply with all the directives of these regulatory bodies. Therefore, how could the company be in violation of the law? Furthermore, it argued that the FCC, not the Justice Department, had the expertise and the Congressional mandate to regulate the telecommunications industry.

At the close of this phase of the trial, which lasted several months, AT&T made a motion for dismissal. Judge Greene denied the request in September saying, "The testimony and the documentary evidence adduced by the government demonstrate that the Bell System had violated the antitrust laws in a number of ways over a lengthy period of time . . . the burden is on the defense to refute the factual showings."

Judge Greene's statement was interpreted by many as meaning he had already reached his decision—"guilty as charged."

It was becoming more and more necessary for AT&T to find some out-of-court settlement. As George Saunders was later to observe, "We were confronted by a judge who wasn't hearing our side of the case. That was the concern."

Judge Greene, some observers felt, was acting like a prosecutor in his questioning of Bell witnesses. Further, he had made it clear he believed that competition in the telecommunications industry would result in the cheapest and best service.

At the beginning of 1982, the settlement came. On January 8, an agreement hammered out between the Department of Justice and AT&T was announced—the Bell System would be dismembered.

The queen was dead. Long live the queen.

8

The Death Knell

"I fear that the breakup of AT&T is potentially the worst thing to happen to our national interests in telecommunications that will ever occur."

—BARRY GOLDWATER, U.S. Senator from Arizona, 1983

It contained only three thousand words, that little document that broke up the largest corporation in the world. Three thousand words and it was remarkably simple considering the complexity of the underlying issues and the tons of paper which went into evidence.

The settlement was an alteration of the 1956 consent decree and was formally called the Modification of Final Judgment (MFJ).

Basically it had nine provisions.

1) The 1956 consent decree was voided.

2) The Bell operating companies were to be totally separated from AT&T along with appropriate facilities and personnel. The companies were allowed some degree of consolidation—the number went from the then 22 to the present seven.

3) The operating companies were required to establish a single point of contact for national emergencies. They were allowed, if they wanted, to share a central staff to work on common problems.

4) Western Electric and Bell Telephone Laboratories could stay part of AT&T.

5) The license contract—the fixed percentage the local companies paid to AT&T for Bell Laboratories and other services—and the Western Electric supply contract were terminated.

6) Exchange service and interexchange service were separated. An exchange was loosely defined as a region containing not more than one Standard Metropolitan Statistical Area. To avoid confusion with prior telephone company usage of "exchange," which took in a much smaller area, the term LATA—Local Access and Transport Area—was coined for this larger area. Service within LATAs would be the province of the local operating companies.

7) The local companies (which predictably came to be called the Baby Bells) could not provide information services, such as cable television, or manufacture equipment.

8) The local companies had to provide equal access to their networks for all interexchange carriers (AT&T, MCI, Sprint, *et al.*) who wished to connect to them.

9) Within six months of Judge Greene's approval, AT&T had to provide a detailed Plan of Reorganization (POR), spelling out precisely how the provisions of the settlement would be implemented. These provisions were to be put into effect no later than eighteen months after Judge Greene's approval of them.

So why did AT&T agree to settle? Why didn't it brazen the suit out, as IBM did. (Ironically, on the same day the AT&T

settlement was announced, IBM's antitrust suit was dismissed.) Why didn't AT&T go the full run of the trial in hopes of getting Judge Greene to see the merits of its case? Or even if the judge ruled against the company, there was the hope of getting the judgment overturned on appeal. And there was always the possibility the Reagan administration or a subsequent one would have eventually folded the Justice Department's tents and dismissed the case.

As it was, many Bell managers thought first reports of the settlement announcement were somebody's idea of a joke. There was total disbelief. Why come so far and then give up?

There were all sorts of explanations. One was Charles Brown's statement that pursuing the case would be long, costly and debilitating with far worse consequences to the company than the negotiated settlement. It should be remembered that Bell wasn't just fighting the Justice Department. The government's antitrust case was one of many. Private companies—notably MCI—had also sued AT&T for alleged anticompetitive or predatory behavior. MCI had already, in 1980, won a judgment of $1.8 billion, the largest such award to date. Later this judgment was partially reversed on appeal.

All in all, there were forty to fifty similar suits in varying stages of investigation, litigation, and appeal. Settlement of the government's case was expected to lead to a resolution of these other actions.

Brown also had internal factors to consider. The ten years of being under the cloud of the antitrust suit had taken a toll on the day-to-day operations of the entire Bell System. Rules and regulations were changing rapidly, complicating every manager's job. It was an all-pervasive presence, hanging over everyone's head and diverting attention from the day-to-day job of furnishing top quality telephone service. Few decisions were being made without considering what influence they might have on THE LAWSUIT. It's nearly impossible to provide top quality service under such circumstances.

Then, too, AT&T was not getting support from where it might have expected some. For instance, the labor unions representing more than half a million Bell System employees took no stand to protect their members' jobs.

And so Brown succumbed.

That's more or less the official view of events. However, one of the authors (Constantine Kraus) has a somewhat different perspective.

It came as a result of a luncheon conversation he had with John deButts at New York's Union League. It was 1979, a little less than a year after deButts retired.

DeButts had to give in, he told Kraus. It was all settled. The Bell Laboratories, Western Electric, and AT&T would stay together. The Long Lines subsidy of the local companies would cease and be replaced by an access charge to customers. And Charles Brown had gotten the chairmanship of AT&T with the proviso he would not fight Washington. It was all wrapped up, all settled.

It was 1979, a little more than a year before the U.S. v. AT&T went to trial.

AT&T was willing to take a loss, as long as that loss was on its terms.

Be that as it may, the official version of events was correct in one respect. Bell System management, distracted by the eroding effects of the FCC decisions and the suit, couldn't help but let other aspects of the business suffer. In particular, Bell Laboratories.

In the past AT&T top management had spent considerable time and energy overseeing the Laboratories' activities. The future was born there. But in the seventies, that time and energy had to be mobilized to meet the threats and assaults of the legislative, legal, and regulatory challenges.

It stands to reason that these distractions impeded progress. Without them, Picturephone service might have been introduced successfully. The network might have been completely digitized five or ten years sooner. Any number of new customer services might have been developed and made available. U.S. telecommunications was undoubtedly set back a number of years. This country has lost its place as the industry's world leader. We're now playing catchup with Europe and Japan.

It didn't have to be that way.

The pressures of the pre-divestiture years were not

confined to Bell's top management. Employees at all levels were wracked by the uncertainty hanging over the system—and therefore hanging over their careers. Thousands had chosen to work for AT&T because of its stability and dedication to service. Suddenly the rules were changed. They were working for a rapidly changing organization with an uncertain future, where profit was sure to ride roughshod over service.

In many other industries, employees disturbed over their company's future start to send out resumes. But despite the perceived turmoil at Bell, tradition won out. Employees had signed on with Bell for life and the concept of jumping ship and getting a job somewhere else was still foreign. Most employees stayed with the company.

Still it wasn't easy. After the settlement it sometimes felt as if not only the rules had changed, so had the entire game. Bell Telephone Laboratories, Western Electric, AT&T, and Bell operating company employees had spent whole careers working together. Organizational lines weren't hard and fast, transfers between the different corporate entities were common and even expected as part of career development, and interorganizational secrets were almost nonexistent.

The one million employees had been united in an enterprise of which they were proud—providing what was undisputably the world's best telephone service while at the same time producing a constant flow of inventions to advance the communications art.

Then bingo! It was all changed. The environment of cooperation and easy access was gone. Employees were branded "Western Electric" or "Bell Operating Company," and they had to remember exactly for whom they worked.

In some buildings shared by two Bell entities, lines were literally drawn on floors to separate personnel and equipment—which in many cases were almost functionally inseparable. AT&T and operating company people who had once communicated freely had to get passes to visit each other. And it was verboten to talk about anything that might give an unfair advantage to potential competitors.

Bell employees were not just concerned about their per-

sonal futures in this tumult. They were even more frustrated
because the dramatic changes in telecommunications policy
and Bell's reaction to them violated every tradition of the
business they had known. Moreover, they violated common
sense.

Not too surprisingly, morale at Bell plummeted. A private
survey conducted in 1981 indicated that morale there was
among the lowest of any major corporation—a stark contrast
to past surveys. While top management gave the situation
much attention prior to divestiture, there's no evidence matters improved before the breakup.

Many employees felt the company they knew, their Ma Bell,
had been brutalized and raped in front of them and they had
been powerless to save her.

But that didn't stop them from taking on the challenge and
mandate of complying with the settlement provisions, with
much of the same spirit they had previously displayed in
times of national emergency or disaster.

While a vast majority of managers vehemently disagreed
with having to dismantle what had been so carefully built,
they accomplished, on schedule, a corporate divorce and asset distribution that many thought impossible.

As Michael Wines said in the *National Journal* in 1982, ". . . a
Dutchman named Jan Asscher poised his chisel in 1908 over
a 1 ¼ pound rock, lifted his jeweler's hammer, and deftly delivered a blow. When he finished, the object of his attack—the
Cullinan diamond—had been reduced to 105 perfectly fac-
eted, glittering gems. . . . The object is too cleave Ma Bell into
at least two corporate giants. . . . The question now is whether
the breakup will produce new corporate gems or a pile of
considerably less valuable shards."

If the jeweler's hammer was to be forced into their hands,
the Bell employees would become Jan Asscher. Thousands of
managers worked long hours for two years on a task most
found odious. But if it was to be done, they wanted to do it
right. As Charles Brown put it. "What had to be done by our
people was done with dispatch, with courage, and with
'class.' "

And so it was, on December 31, 1983, the books were closed on the Bell System and Ma Bell, as the world had known her, breathed her last.

Divestiture was complete.

9

A Tale of Two Judges

"My obligation was to apply the Sherman Antitrust Act."
—JUDGE HAROLD H. GREENE, 1985

"What the FCC actually did over the years was talk about competition (in reality contrived) in order to achieve deregulation ... contrary to the best interest of millions of Americans ... with a loss of service, quality and higher costs to those less able to afford this essential service."
—JUDGE CHARLES R. RICHEY, 1982

It was the luck of the draw.

The AT&T antitrust suit could have gone to Judge Charles R. Richey.

When Judge Waddy died and his caseload was distributed, the U.S. vs. AT&T went to Greene and another antitrust case, Southern Pacific Communications Co. vs. AT&T, was assigned to Richey.

Even though the Southern Pacific case alleged many of the same things as the Justice Department's, the conclusions and attitudes of the two judges were worlds apart.

On the one hand there was Judge Greene who showed from the outset a decided bias against AT&T. Even Assistant Attorney General Baxter admitted that Greene's written opinions revealed hostility toward the company. What became paramount was AT&T's perception of Greene's bias. That led the company to resolve the dispute out of court.

The result was that Judge Greene, practically overnight, became the nation's most powerful figure in the telecommunications field. Aided by only two law clerks, he made final decisions involving the restructuring of a technically complex industry. His decisions covered everything from billions of dollars in investments to the name—Bell Communications Research or Bellcore—for the company that would do technical development work for the Baby Bells.

But Judge Greene did far more than change the corporate setup of AT&T. He caused the entire telecommunications network to be rebuilt physically by separating the exchange and interexchange services. While not wishing to question the judge's intelligence, one has to wonder about his technical expertise. Anyone with even a little knowledge of communications engineering knows that this separation was illogical and wasteful.

Here was a network designed for maximum efficiency by generations of scientists and engineers—and whoosh! Someone with a law degree from George Washington University pulls it apart in two years and makes it far less efficient. Small wonder Bell System engineers and managers were frustrated.

But it has to be assumed Judge Greene had no idea the negative impact his decision would have on the nation's telecommunications network.

And then there was Judge Richey and the Southern Pacific Communications Corporation case which got underway several months after the settlement in Judge Greene's nearby court.

Southern Pacific was Sprint's parent company. The charges it brought against AT&T were almost identical to MCI's and the government's. AT&T's attorney Saunders made many of the same arguments, used the same witnesses, the same evidence as had been used in the other cases. Sprint was a "creamskimming operation," he argued.

But what a difference in outcomes. Judge Richey dismissed the case—with prejudice, costs in favor of AT&T. He based his dismissal on several significant points, finding—

1) That there was no evidence of anti-competitive or predatory behavior by AT&T and, further, Southern Pacific had not suffered injury resulting from any AT&T actions.

2) That matters subject to public utility regulation should not be subject to antitrust laws, that these laws had not been designed to destroy an essential public utility.

3) That there had been impropriety on the part of some major Southern Pacific witnesses. One had improperly removed files from the FCC. Several had worked as consultants for AT&T competitors—MCI and Datran—while on the FCC payroll. That these witnesses were out to wreck AT&T, and their testimony was worthless and their actions gave government a bad name.

4) That AT&T had made all reasonable efforts to comply with what it understood to be FCC requirements, but the FCC had an institutional bias against providing guidance to AT&T on the procedures it wanted followed. Further, AT&T may have been deliberately misled by FCC directives.

5) That there existed a blatant conflict of interest in the FCC's Common Carrier Bureau, and the conduct of the FCC staff had been unprofessional and inadequate in allowing contrived competition.

6) That the FCC had usurped jurisdiction over intrastate matters, perhaps contrary to the public's interest. Further, that the FCC had interpreted the public's interest as being what certain powerful individuals wanted rather than what the public needed—safe, reliable, affordable service.

"The FCC ... put the Specialized Common Carriers into business for the benefit of a few without the taking into account the myriad of problems to our people and to our national security," Judge Richey wrote.

Richey ended his decision by quoting a speech made by FCC Commissioner Benjamin Hooks in 1973.

"So I say there is one thing I'm sure of, we do have a basically good communications system. That doesn't mean that it can't be improved both by the company itself, and certainly by some of the things that we regulatory folk at both the FCC and the state level are doing.

"But above all let's remember that our telephone communications system, strong as it may seem—may be as fragile as Humpty Dumpty. If we mess with it too much we might topple it off the wall, then all the king's men may not be able to put it together again. All talk then of regulation versus free market, will it mix, or will it not, will be moot."

And the gavel came down in that courtroom.

Two judges, two different attitudes, one nation's telecommunications network in disarray.

How could the two men, with the same evidence, come to such different conclusions? If AT&T was so successful in getting its points across in one courtroom, why hadn't it gone the full route in another? And why wasn't Richey's decision given more publicity?

Webster's defines "conspiracy" as "a planning and acting together secretly, especially for an unlawful or harmful purpose, such as murder or treason."

One thing is for sure. There was a murder and Ma Bell the victim. Beyond that, little at this point is conclusive. Still while sifting through the evidence, one can't help detecting an unpleasant odor and wondering if it isn't the stench of conspiracy.

10

The Damage Report: Ma Bell and Her Orphans

Winds die down. Clouds roll away. Water begins to subside. And survivors push away debris to assess the damage.

Enough time has passed since the settlement in 1982 and its implementation two years later to do some tallying. What has happened to AT&T, the new Baby Bells, their employees, the other common carriers, the independent companies, the nation's economy, its defense, and the public, we the people who use the phone system?

How much was lost? Did anyone come out ahead? Did a better telecommunications system rise out of the rubble? And then there's the bottom line—what did this unnatural disaster cost, in dollars and cents, for now and the future? It's a staggering sum, when all elements are added up, and the country will be paying it off for generations to come.

But first, the individual players and how they fared.

Ma Bell. She was sacrificed and from the one mammoth company were cleaved eight large ones. Her uniqueness—

unifying the organization that was producing technology with those that were applying it to customer service—was lost.

The parent company, AT&T, was left childless with the twenty-four local operating companies torn from its corporate bosom. Left at AT&T were the long distance business and the research, development, and manufacture arm, AT&T-Technologies, which is a combination of the former Bell Telephone Laboratories and the Western Electric Company. AT&T was also saddled with a totally unfamiliar job, namely providing customer premises equipment in competition with all the other suppliers. Under the old Bell System, the local operating companies had handled customer equipment.

The two local operating companies—Southern New England and Cincinnati and Suburban—that were not solely owned by AT&T now operate independently. The other 22 were combined to form seven new companies.

New England Telephone and Telegraph Co. New York Telephone Co.	NYNEX Corp.
New Jersey Bell Telephone Co. Bell Telephone Co. of Pennsylvania Diamond State Telephone Company Chesapeake & Potomac Telephone Company of Maryland Chesapeake & Potomac Telephone Company of Virginia Chesapeake & Potomac Telephone Company of W. Virginia Chesapeake & Potomac Telephone Company of Washington	Bell Atlantic

Ohio Bell Telephone Company Indiana Bell Telephone Company Illinois Bell Telephone Company Michigan Bell Telephone Company Wisconsin Telephone Company	Ameritech Corp.
Southern Bell Telephone & Telegraph Co. South Central Bell Telephone Co.	Bell South Corp.
Southwestern Bell Telephone Co.	Southwestern Bell Corp

Northwestern Bell Telephone Co.
Mountain States Telephone
    & Telegraph Co.
Pacific Northwest Bell Telephone
    Co.
	U.S. West Corp.

Pacific Telephone
    and Telegraph Co.
Bell Telephone Company
    of Nevada
	Pacific Telesis Corp.

William Baxter in 1985 claimed, "I knew how to break up the company without damaging it too much. If you do a divestiture, you have to create viable commercial enterprises."

Whether one agrees with his "not too much damage" profession, it has to be acknowledged that the resulting seven Baby Bells in their first years have done well financially.

Their share prices on the New York Stock Exchange doubled. Several have had stock splits, much to the benefit of investors who had gotten in on the bottom floor.

The stock prices reflected the rosy earnings pictures of the companies, most of which have had record highs—embarrassingly so in some cases. William Baxter can't take credit for these increased revenues, however. Most of them were due to rate increases, authorized by state public utility commissions on orders from the FCC. The commission had to offset the subsidy losses from AT&T's long distance revenues resulting directly from long distance competition. Actually the state public utility commissions should have reduced local rates to compensate customers for having to buy their own equipment and having to pay for maintaining it, thereby lowering the operating costs of the local companies. Instead rates were increased an incredible $35 billion annually. This, in effect, overcharged the customer $65 billion annually—that's the cost of supplying his own equipment plus the rate increases.

The locals were also able to raise earnings through austerity measures. And then, too, modernization projects that had been started long before divestiture finally began paying off.

Without many of the old strictures of a regulated monopoly, the new companies have embarked on new ventures. The Modified Final Judgment contained no language prohibiting them from moving into new areas as long as they did not use their monopoly power to impede competition and did not subsidize these new ventures with money from their regulated activities. Pacific Telesis has actually reversed this. It announced that when it starts making profits with its competitive enterprises, it will subsidize the regulated portion of its business with those profits.

Originally Judge Greene, as part of the settlement, had to okay these new ventures. But in 1987, the prior approval requirement was dropped and the companies, for the most part, became free to run where they wished. And they have been running to the tune of $5 billion in new acquisitions.

Take Bell Atlantic. Besides its regulated telephone services in six states and the District of Columbia, to date it operates these subsidiaries:

Bell Atlanticom—markets, installs, and maintains telecommunication systems and equipment.

A Beeper Company—sells paging equipment nationwide.

Bell Atlantic Mobile Systems—markets cellular mobile radio products in the mid-Atlantic region.

Bell Atlantic Properties—develops, invests, and does consulting work on commercial and light industrial real estate.

MAI Canada—markets and maintains computers and provides custom software throughout Canada.

Sorbus—a nationwide computer maintenance operation.

Telecommunications Specialists—rents, installs, and maintains communications systems in Texas.

Bell Atlantic TriCon Leasing—industrial equipment leasing and financing throughout the United States and Canada.

Bell Atlantic Technical Ventures—develops computer software as a joint venture with various universities.

Bell Atlantic Business Supplies—markets paper, ribbons, diskettes, and other supplies to computer users.

Bell Atlantic International—provides telecommunications network consulting services to overseas customers.

Electronic Service Specialists, Ltd.—provides parts and repairs for Digital Equipment Corporation computer equipment.

Technology Concepts, Inc.—provides computer software design and development services.

Some Bell operating companies are planning to manufacture communications equipment overseas, but not because they wouldn't like to domestically. The settlement agreement prohibits them from doing that. What's more, they may not import what they manufacture abroad. Which boils down to the local companies not being allowed to use what they make. Must make sense to somebody.

Other operating companies have gone much farther afield. U.S. West has stretched into the securities brokerage business and consumer electronics.

All this branching out, however, has not proved terribly successful. In February, 1988, Bell Atlantic announced it was throwing in the towel on its stab at computer retailing. Two years earlier it had purchased the 65-store CompuShop chain. It decided to sell the sixteen shops that remained to the New Jersey firm, CompuCom Systems Inc.

"The market never really did develop," a Bell Atlantic spokesman explained.

The regulated business accounts for approximately 92 percent of revenues. In 1986 that added up to $62 billion out of $67 billion for the seven Baby Bells.

Unofficial statistics quoted by the North American Telecommunications Association indicates a total loss of $471 million in 1985 by the seven operating companies from their little forays into competitive subsidiaries. During that same period, the regulated services showed profits of $12.2 billion. It may be quite a while before Pacific Telesis' competitive subsidiaries start forking money over to the regulated end of the company.

Meanwhile back at AT&T.

In 1986, the company had to write off $1.7 billion against earnings, reflecting poor computer and equipment sales. On the plus side, even though earnings and profit margins were shrinking, its customer base is expanding. Also, AT&T has regained its market share lead over competitors in most categories of telephone equipment sales. Still, sales costs remain higher than for most of their competitors, and the best bet is there will be further personnel cuts along with other cost reduction measures.

The new AT&T is a smaller, leaner company with much wider horizons in terms of where and how it can operate. Once a strictly "Made in the USA" company, it has expanded its overseas sales and services operations. It also imports foreign-made goods for sale and use in the United States. This includes not only materials and systems for the country's long distance networks, but also items sold directly to the public, as is the case with AT&T computers manufactured by Olivetti in Italy.

Regrettably, one of the worst casualties of divestiture has to be at the former Bell Telephone Laboratories. True, the Laboratories still exist as part of AT&T Technologies, but it's not the same. Since divestiture, in its new guise it is just another industrial lab.

The weaker, smaller post-divestiture AT&T can no longer afford the free-spirited and wide-ranging investigations in which Bell Laboratories scientists once engaged. Basic research will have to be deemphasized as the profit statement takes priority over quality of service. More and more of the Laboratories activities will be directed toward specific corporate goals.

As was stated in the Washington *Post* in 1982, "Competition is good for commercial efficiency and for product development—but not for scientific research."

There is little doubt that the old Bell Telephone Laboratories, often called a great national resource, has been dealt a severe if not fatal blow.

The nature of research and development has changed in another, insidious way. In the old days, AT&T had to share the technological and scientific fruits of its own research. Bell Telephone Laboratories created the information age, but was forbidden to pursue its technology except within a narrow field. This is no longer the case and the free ride is over for those who built their businesses on Bell technology.

For the corporation that's good. Had AT&T been allowed to follow its technology when its scientists invented the transistor, the company would probably dominate the computer and microelectronic industries today.

But for the rest of the country, it may not be so good. AT&T scientists and engineers can no longer openly discuss what they're working on or their results with colleagues in other organizations. Their work will become increasingly more clandestine and directed toward specific and shorter term goals.

The sad reality is scientific research flowers and flourishes best in a free and open environment. Overall progress can be severely impeded when information is not shared.

Of course, under the new arrangement of divestiture, these outside competitive organizations include the local operating companies. Bell Telephone Laboratories is structurally separated from them, and the previous relationship between them broken. However, at divestiture some former Bell Laboratories scientists were reassigned to Bellcore, the organization that is the single point of contact for national emergencies and the shared staff to work on common problems. These scientists, naturally, remain accessible to the Baby Bells. Actually the former Bell Laboratories split its personnel after divestiture in three ways—15 percent going to Bellcore, 18 percent to the AT&T division responsible for products used on the customer's premises, and 67 percent remaining with AT&T-Bell Laboratories.

Another divestiture casualty, at least in terms of effectiveness, was the Technical Assistance Centers which had been established by Western Electric. These centers maintained huge databases with information on electronic switching system failures, problems, and solutions. These databases were tapped into by Bell Laboratories and the local companies, allowing the latter to quickly diagnose and correct problems. Similar information is still available, but it is not centralized. Rather, it is divided among a number of manufacturers. Also, the databases are only as complete as the local operating companies choose to make them. If they don't supply the information, the information doesn't get stored. As a result this tool is considerably less effective than it had been.

When assessing the damage inflicted by divestiture, another in no way insubstantial item must be considered. The efficiency of the network.

In choosing to tear apart the finely tuned, interconnecting organization providing long distance and local service, Judge Greene chose an incredible illogical—and expensive—reassignment of responsibility.

Under his arrangement, the Bell System telephone service areas in the United States were divided into some 160 segments, those Local Access and Transport Areas (LATAs).

Service within each LATA went to the appropriate Baby Bell. Inter-LATA service was to be handled by AT&T and any other carrier who wished to enter the field.

This structure, along with the rules prohibiting joint ownership of facilities, has led to some incredible network costs and inefficiencies.

Consider the situation in downtown Pittsburgh where there are two adjacent buildings that were once both owned by the Bell Telephone Company of Pennsylvania. But some of the equipment in them had belonged to AT&T, or to both companies with ownership ratios related to usage. After divestiture, this arrangement was no longer acceptable. So each company took a building, but there were very large amounts of equipment in the buildings still owned by the other company or by both. If the equipment wasn't of use to the owner of each building, eventually it would have to be scrapped. In the case of a jointly owned piece, the company whose building it was in got it and the other company had to build a new one, even though the existing equipment would be operating well below capacity.

For example, in the building that went to AT&T was a No. 4E digital electronic switching machine that had been handling nearly all long distance traffic into and out of Pittsburgh. That was fine until joint ownership was prohibited. This meant that Bell of Pennsylvania had to build a new $6 million digital electronic switching machine and then spend another $1 million to transfer circuits from the No. 4E to the new machine. Today both machines are operating well below capacity, furnishing separately and at higher cost service that had been provided adequately on a single switcher. This scenario was repeated throughout the United States.

But that wasn't all. Before divestiture, the integrated nationwide network handled all offered traffic in a unified and efficient way. If the most direct route was tied up, a call would be sent through the next best route. But then something called the POP concept came into existence. Under it, the local operating companies are only allowed to interconnect with the inter-LATA carriers at designated "points of presence"—right, POPs.

So take the case of a call going from Philadelphia to Lancaster, Pennsylvania, some 50 miles away. Before divestiture a multimillion dollar digital microwave route had been built by Pennsylvania Bell to connect them directly. At that time, after all, the local company handled all intrastate telephone service.

But wait, now it's postdivestiture. A call between Philadelphia and Lancaster is inter-LATA. AT&T is handling it. But AT&T's POP for the LATA containing Lancaster is Harrisburg. What this means is that now a Lancaster to Philadelphia call goes three times the distance, first up to Harrisburg where AT&T picks it up. Service at best was unaffected, but the cost was substantially increased. Multiply that situation by hundreds, probably thousands of locations across the country, and the additional costs are up there in the billions.

The POP concept has had another adverse effect on the network. The unified intercity network, developed and built over many years, was hierarchical in nature and designed to operate so that there were many routings available between any two points. So-called "high usage" trunk groups were engineered to carry 80 percent of peak hour calls. Calls it couldn't handle overflowed into a series of less direct high usage groups. These in turn overflowed traffic onto a route of last resort, a final route.

High usage facilities were usually fairly direct and inexpensive. Final routings, handling only a small fraction of the total traffic, were more roundabout and costly, with intermediate switching. This efficient network of high usage/final trunking assured that more than 99 percent of the offered traffic would be completed, with a high degree of protection against overloads and network failure.

This arrangement has now been partly dismantled since all inter-LATA calls must pass through an interexchange carrier's POP. If the route to the POP, or the POP switch itself, is overloaded or encounters a failure situation, too bad. There are no alternatives and the call is blocked. So the POP concept has created less efficient service at higher costs.

Added expense is piled onto added expense under the nonsensical LATA separation with no additional benefit to the customer. Another area where money has been wasted needlessly is with operator services.

The bulk of intercity telephone traffic is dialed directly by the customer with no operator intervention. However, in an emergency or when dialing, billing, or transmission problems occur, the customer is accustomed to dialing "0" for assistance. In the past, the great majority of operators answering the "0" summons were local operating company employees, and they handled AT&T calls as well.

As part of the divestiture settlement, most of these operators went on the AT&T payroll. When they provided local assistance or completed intra-LATA calls—local company responsibilities—AT&T billed the Baby Bells for the services. After a while, for the sake of public relations and lower costs, the local operating companies decided they wanted their own operators dealing with the local traffic. In many cases new operating centers had to be built at considerable expense and operators who had been transferred had to be reemployed, leaving AT&T with excessive facilities and equipment.

Competition in telecommunications was a poor idea. Need an additional example? Look at cellular mobile radiotelephone service. After delaying its introduction for fifteen years—permitting numerous European countries to move far ahead in this field, even though the technology was developed by the Bell Laboratories—the FCC decided the service should be competitive.

Two entrants would be licensed to serve each geographical area. That meant, in most cases, that one of the two would be the local telephone company. But wait. The FCC didn't want the telephone company to have any unfair advantage. What to do? What the FCC did was order the telephone company to furnish the service through a separate subsidiary.

To guarantee this separation, barriers to communication and cooperation between the telephone cellular subsidiary and the local telephone business were erected. In Pittsburgh, for example, Bell Atlantic Mobile Systems—BAMS—constructed a multimillion dollar No. 1A electronic switching system in a rented building a few blocks from Bell's main telephone building downtown. Of course, the Bell building already had several No. 1A systems with more than enough idle capacity to take care of BAMS's needs. But no matter, separation requirements said BAMS couldn't use Bell equipment, and that was that, even though, in terms of call switching capacity, the new BAMS facility was totally superfluous.

What's a few wasted millions of dollars between friends? The FCC wanted contrived competition, then the consumer will pay for it, because, of course, all this large unneeded expenditure meant unnecessarily high rates for cellular radio service. Small wonder that our development of this service has lagged behind that of many other nations.

AT&T and the Baby Bells are competitors now. And it has become "all-out warfare," according to one longtime industry consultant.

Nor should it be forgotten that the Baby Bells are competing with each other. The *Wall Street Journal* has even called them "seven John McEnroes," because of their cantankerous bickering. They're invading each other's territories, disparaging each other's integrity, looking for ways to undercut each other's efforts.

According to one Washington-based industry consultant, "Relationships have broken down. People aren't talking to each other" to the extent that former friends "can't stand to be in the same room with each other."

Divestiture was more than pulling apart a company, printing new letterheads, opening up competitive markets. It was also people. Divestiture meant real day-to-day, year-to-year changes for the former Bell System employees.

Few of them had felt that changing the industry was going to benefit them personally. Actually, a number of individuals did benefit by faster advancement than they would have experienced under the old system. And some employees feel quite comfortable in today's more competitive environment. "AT&T is no longer the warm parent," one NYNEX executive said in 1986. "Now it's another cold competitor."

Gone are the days when one million people worked hard and long in an atmosphere of cooperation, where the foremost goal was to provide the best service, when doing a good job wasn't measured in profits alone.

But that is to the liking of some employees. They cite a greater sense of proprietorship, and the opportunity to more directly affect the company's bottom line. This desire to increase profits is being encouraged by most of the Baby Bells through a variety of profit-sharing plans, where 20 percent or more of an employee's salary may be based on specific results of the company, the department, and the individual. In the marketing arms, more than half an employee's compensation may be linked to sales performance.

One by-product of divestiture was the creation of more top-level executive positions since there were now eight independent organizations instead of a single corporate identity. Those in these new top positions have fared quite well. In 1986, the chief executive officers at AT&T and the seven local operating companies each earned, on average, in excess of one million dollars in salaries and bonuses. No top Bell System officers came close to that in predivestiture days.

Another benefit for some employees was the increase of opportunities outside the ex-Bell System. Those conversant with the workings of packet switching, fiber optics, artificial intelligence, software design, and other related fields now have far more opportunities for profitable employment, either with the former Bell System companies or their competitors.

But other individuals got blasted by divestiture and the

earlier FCC actions. About sixty thousand jobs were eliminated from AT&T's Information Services division, largely because it lost a good chunk of the customer premises equipment market to foreign and domestic competition.

Another eleven thousand jobs were eliminated at AT&T Technologies, mostly among those who had worked at Western Electric. At the local operating companies, more than thirty-five thousand jobs were cut, for cost-reduction reasons and because of the virtual elimination of the installers and repairmen who used to work on the customers' premises—work now done by independent contractors or by the customers themselves.

These job cutbacks were softened somewhat by employees who were encouraged to take early retirement through a variety of incentives including enhanced pensions or termination allowances of up to a year's salary. Where these incentives did not reduce the employee roster enough, transfers were offered. Only as a final resort were layoffs implemented. Those who survived and remained now have more challenging jobs, more closely related to performance. But along with that comes much less job security.

All this change and uncertainty took its toll physically on some employees. Psychological studies of several hundred executives were conducted. It was found that the largest change in the history of the corporate world had created stress and even physical illness. Besides the upheaval, there had been the long hours logged in to bring about the dismantling of the Bell System. Following that, there had been a larger than usual number of job transfers, many of which entailed a home relocation. It was a hard, upsetting time.

Over at the Bell Telephone Laboratories, scientists were also uncertain. Most felt they would lose some of their traditional freedom. They wondered if working for the new AT&T would be as rewarding as for the old. Other companies, hoping to benefit from the uneasiness, moved in with tempting job offers for thousands of scientists. In the end, hundreds left.

For the employees, then, divestiture was a mixed bag. Talk to the San Francisco cabbie who once thought he had a lifetime job as a telephone company installer, and there will be one view. Talk to the executive with what he perceives as a more challenging job with a paycheck he knows is a lot fatter, and there will be an entirely different perspective.

The only thing definite is nothing stayed the same.

11

The Damage Report: The Other Common Carriers

Divestiture might have started champagne corks popping at the offices of AT&T's competitors in the long distance market, the other common carriers—or OCCs for short—but lately the bubbly has gone flat.

Originally it was carved in stone—Long Distance Service *was* AT&T. Then in 1977, after years of maneuvering, lobbying, testifying, and then pretty much ignoring the FCC, MCI opened the floodgates for non-Bell long distance service with its Execunet service—although its President Bill McGowan had earlier assured the FCC his company had no intent or desire to compete directly with AT&T.

After that, customers had an ever-widening choice. It was no longer AT&T or write a letter. Many chose AT&T's competitors. And why not? It seemed a fairly simple choice. These other carriers were undercutting Ma Bell's rates—and at the same time were reaping a substantial profit.

This was no sleight of hand by the competition. First off, long distance rates of the former Bell System had been set at a level well above the cost of providing the service. The long distance end of its business was extremely profitable—why else would hundreds of companies have been scrambling to get a piece of the action?

Bell was permitted this overcharging in long distance because the excess profits offset the losses in the local exchange through a process called separations payments. This subsidy amounted to more than $10 billion a year.

The rationale was that as long as the overall enterprise earned no more than a fair rate of return, the subsidy was permitted for the social good. It was a question of the rich—the business customers and more affluent residence subscribers who were heavy users of long distance—helping the poor—the local exchange customer who made few long distance calls. This bit of social engineering on the part of the Bell System had long been condoned and even encouraged by Congress and the regulatory agencies.

Enter the competition, which could set its own rates. They weren't saddled with the higher rates that AT&T had and that AT&T could not arbitrarily lower.

Another reason for the other companies' rate advantage was the fact that they could select their markets. Not being self-destructive, they went only into the more profitable ones where service could be provided more cheaply. And they didn't have to spend money on emergency backup facilities, alternate routes, or network controls. But most important, only AT&T had to subsidize local exchange telephone service to the tune of forty cents on every long distance dollar earned, while the MCIs paid almost nothing. Adding this up, the FCC, in its ill-conceived attempt to create competition where competition doesn't belong, has given the new carriers on the order of $7 billion. It was easy for MCI and the others to charge less than AT&T for a call between New York and Los Angeles.

But the sad fact of economic life is the public never gets something for nothing. Those calls being made on MCI meant fewer long distance calls on AT&T. Fewer long distance calls on AT&T equalled less money to subsidize local exchange service. That inevitably led to a need for higher local rates. No matter how one looks at it, a long distance call starts out with a telephone. To reap the cheaper rates for that crosscountry call, the consumer ended up paying more each month to have a phone in his house. In a very real sense, the profits of the OCCs came directly out of the pockets of all local exchange customers, not just their own customers.

AT&T's share of the inter-LATA business—in the early 1980s a $60 billion market—dropped to less than 85 percent with the introduction of competition. It still could offer more reliable service, generally better transmission, universal access, and easier dialing—fewer digits needed to complete a call—but that didn't offset the lure of the OCCs' lower rates for some customers. A 1987 Paine-Webber survey showed that 65 percent of the customers who went to another common carrier did so because of price.

All looked black ink and great prospects for the OCCs. But then the rules were changed. Something called equal access, mandated by the Modified Final Judgment as part of the divestiture agreement. Equal access meant that the local operating companies had to provide the OCCs the same quality access to their networks as had been furnished to AT&T.

Before equal access, which began to be available in 1984 and was essentially completed at the end of 1987, the customers of the other common carriers had to dial up to twenty-two digits to make a long distance call. Once equal access was provided, hallelujah, there was no need to grow old making a call. Only eleven digits—the same as for AT&T service—were needed. It might seem that equal access would only increase the desirability of signing on with a Sprint or MCI. Equal access was more convenient, and it improved the quality of transmission. Why would anyone stay with AT&T and pay more for the same service?

The costs to the Baby Bells for making modifications to provide equal access were astronomical. Each central office had to be modified to permit "1 + 10 digit" dialing for access to the interexchange carrier of the individual customer's choice. The customer would select carriers other than his primary one with a five-digit access code instead of the "1."

It cost the New York Telephone Company alone somewhere around $2.5 billion to provide equal access. This is $300 per customer. Such costs are factored into the overall rate base and inevitably paid for by the customer. Projecting the New York estimate nationwide, equal access costs exceed $20 billion. In all fairness, it should be noted that a portion of this cost—possibly as much as half—includes plant modernization that would have been done eventually.

Equal access modifications create the additional inefficiencies of multiple access line groups from each central office or tandem office to each intercity's POP, instead of the former single efficient access circuit group. Again, economies of scale were thrown away. All these expenditures contribute nothing to improved service. They merely provide the means to artificially introduce competition into the network.

Beyond the equipment and facility costs, the administrative expenses of maintaining the complex records required must be considered as well as the cost of all those mailings notifying one hundred million customers that they were going to have to make a selection of a primary long distance carrier.

AT&T had equal access cost penalties as well. It has estimated the cost of rearranging its facilities at $2.7 billion. But there's some justice in the world. Even the other operating companies paid a penalty. Because of the short implementation period for equal access as mandated by the settlement agreement and the subsequent plan of reorganization, the OCCs didn't have adequate time to plan their facilities. Consequently the long-range goal of access by OCCs to Baby Bell end offices through centralized tandem offices had to be supplemented with direct OCC end office connections. This forced the OCCs to spend their money twice—first to interconnect to individual offices and later to arrange their traffic into a more efficient tandem pattern.

There was a second catch to equal access for the OCCs. Up until then, the OCCs had paid far less than AT&T did to connect with the local Bell companies. But once the quality of service from the local operating companies became the same for AT&T and the OCCs, MCI and the others lost some of their bargain price connections. With these higher costs of operation, many of the original three hundred or so OCCs went out of business or consolidated—such a merger created U.S. Sprint. The survivors have been forced to raise their rates, making it more difficult to compete with AT&T.

But even with equal access, AT&T still paid more to the local operating companies for connections than the OCCs. Instead of equalizing these charges, the FCC has considered proposals that would permanently give the OCCs the advantage of lower charges—a figure of 35 percent below AT&T's has been mentioned. In essence, the FCC is not creating the free enterprise it has been espousing. It's still giving the OCCs a free ride at the expense of the local exchange customers. The contribution of this subsidization of the OCCs is not only economically wrong, it is totally dishonest. The local exchange customer is not voluntarily, with full knowledge, forking over money to the OCCs. That money is being slipped out of his pocket without his permission. That sounds like stealing.

Another factor has increased the OCCs' difficulties. The FCC mandated a customer line access charge that has been added to each customer's bill. This money goes to the local operating companies, thereby in part taking the place of the subsidies AT&T had kicked in from its long distance revenues. This has meant AT&T could lower its rates, to the tune of more than 35 percent. The chasm between AT&T and OCC rates is now virtually eliminated.

With the incentive of lower prices reduced, some OCC customers have gone back to AT&T, citing poor transmission, excessive cutoffs, billing inaccuracies, objectionable echo and delay in satellite circuits as reasons. Harry Newton, publisher of various telecommunications journals and widely regarded as an industry sage, has said, "The American public is waking up to the fact that a lot of long distance service is pretty bloody awful."

Business customers, with large private line network needs such as centralized airlines reservations offices, are the mainstay of the OCCs' business. But even here, a shift back to AT&T has started. The reason? Better quality of service at AT&T. Better quality that has been measured in studies such as the one conducted by *Data Communications* magazine in August, 1986. It tested data communications links over AT&T's lines versus those of five competitors. It found AT&T's lines were consistently set up faster and performed more accurately.

At the same time this is going on, MCI and Sprint are paying hand over fist for new high capacity facilities, facilities that will, in fact, duplicate each other's and AT&T's. Sprint is now constructing a $2 billion, 23,000-mile fiber optic network interconnecting all major United States cities. (Any time Sprint's commercials flash across the television screen boasting of this great new network, the viewer should scream in outrage. The commercials can only serve as reminders of the excess costs being perpetrated by competition in the telecommunications industry.)

MCI is building its own 10,000-mile network while AT&T has a 24,000-mile one in the works. This is in addition to the many thousands of miles of microwave radio, coaxial cable, and fiber optics routes already serving these same points. The OCCs are spending this money to increase their shares in the long distance market.

This is big money at a time when the OCCs' party is turning into a wake. U.S. Sprint reported losses of more than $500 million in the second quarter of 1987 alone, resulting in a top management shake-up. It also has been beset by billing errors and delays (at one point in 1987, billing was more than fifty days behind schedule). It's been the victim of computer "hackers" who illegally access its network, and computer errors that caused an accidental cutoff in service to twenty-five thousand customers.

Meanwhile, MCI is in worse shape than Sprint and is struggling to hold its market share. It's staying alive on revenues from sources other than long-distance service. In March of 1988, Paine-Webber reported that "Competitively, Sprint is cleaning MCI's clock."

12

The Damage Report:
The Independents

While the settlement judgment was directed solely at AT&T, the single defendant in the antitrust suit, it was to have a profound and possibly deadly effect on the nation's independently owned telephone companies.

For most Americans, before the introduction of MCI and the like, to say "telephone" was to say Bell. However, for people who live in Kinderhook, New York, it's Berkshire Telephone, and in Erie, Pennsylvania, the telephone company is General Telephone. In fact, in the neighborhood of 20 percent of the nation's telephone customers are served by more than twelve hundred independently owned telephone companies (down from about seven thousand in the 1920s) with ten thousand exchanges. Independents operate in every state except Delaware, and their service areas cover about 60 percent of the country's land area.

Most are in rural areas, but cities as large as Tampa, Florida, and Rochester, New York, also are served by independent companies. Some of the independents are very good, on a par with Bell, others so-so, but then there are those that make the customers think of communicating by carrier pigeon. (One explanation given for the abominable performance of the last group is that for years they were not permitted to buy telephone equipment from the best supplier—Western Electric.)

At one time, independent companies and Bell vied for business in the same areas, and customers had to choose. Not any more, fortunately. The Keystone Telephone Company in Philadelphia, which closed up shop just after World War II, was the last to have an overlapping area with Bell. Since then independents operated only where Bell did not, and were interconnected with the Bell System as provided for in the 1913 Kingsbury Commitment.

While the independent companies were never part of the Bell System, before divestiture they generally were operated much as if they were. AT&T set standards, not just for the Bell System, but for the entire industry. The unified nature of the network included the independent companies' portions almost as if they had been integral with the Bell System. This unification was accepted by the independents because of the large subsidies they received from AT&T's long distance revenues.

The independent companies, at least the smaller ones, used the Bell System's resources for technical planning and staff help. They were not usually charged for these services since it was desirable for Bell to provide them in the interest of system uniformity and connectivity. But occasionally that made for an uneasy relationship between Bell and the other companies. Bell was concerned with the "industry solution" to problems, that is, the course of action that would produce the best overall cost and service result, regardless of how it might affect the fortunes of an individual company. In theory the independents subscribed to this philosophy also, but it sometimes conflicted with the concerns for a company's bottom line.

For example, in 1965 a small Pennsylvania independent proposed establishing its own operator location to serve its five central offices rather than continuing to use the large Bell operator switchboard installation in nearby Harrisburg.

It made sense for the independent. Studies had shown it would cost the company somewhat less to handle its own traffic rather than to pay Bell to do it. Also, it would be able to finance building the new facility with a 2 percent loan from the federal government.

Unfortunately for the independent, the proposal did not make sense from an overall industry point of view. Bell's Harrisburg office was much more efficient because of its size, especially during light traffic hours. Furthermore, the proposal meant creating something that already existed, a duplication of services.

There was never any doubt that the independent, once advised by Bell to restudy the proposal, would drop it.

But if the independents gave up some autonomy in the arrangements with Bell, they gained far more. The free technical assistance provided by Bell to all but the largest independents meant the smaller companies didn't have to hire their own staffs. Bell performed fundamental planning studies, made transmission measurements, and provided other similar services. This was a great financial boon to the independents.

Also, recognizing that the independents had no way to expand, being surrounded by Bell and not having access to the more lucrative long distance business, Bell paid them what amounted to a subsidy which was in line with what it paid its own local operating companies.

Of course, that was before the FCC's rulings and divestiture. Now the network is fragmented and there is no overall system planning. The days of free help from Bell are out the window. The independents will now have to hire additional staff or pay consultants to perform the studies and analyses that Bell used to do, or eliminate them completely.

Naturally this will lead to increased costs. Independent company revenues came primarily from local exchange service and toll settlements (the subsidies from Bell).

The smaller percentage had come from the local exchange service, and that was the source least affected directly by divestiture—especially for those companies whose territories are largely rural. But the toll settlement revenue got socked. Toll settlements have been drastically reduced or eliminated. When AT&T got out of the local exchange business, it also got out of the subsidy business. With the deregulation of long distance service and the introduction of competition, subsidies and toll settlements would become history.

To replace some of these losses, the National Exchange Carrier Association (NECA) under the direction of the FCC established a pool of funds. Money flows into the pool from fees paid by inter-exchange carriers for local access. About 55 percent of these fees go into the NECA pool.

The local exchange companies draw from the pool based on their costs of providing service. This means that the smaller, less efficient companies get more from the pool. Several of the Baby Bells don't like this arrangement and are challenging the legality of the whole concept. It appears that the arrangement will be eliminated starting in 1989.

But what has really kept the independents afloat is the fact that the local public utilities commissions have dramatically raised exchange rates, which now include the customer access line charges. They have been able to do this without major public protests because most customers don't stop to consider the fact that even though their monthly local service bills may not be a great deal higher than the past, these bills no longer include equipment and wiring on the customers' premises.

Between the higher exchange rates and the NECA contributions, some of the independents have become embarrassingly prosperous, and one of their major problems is concealing the high rates of return they are getting.

Looking ahead, however, the road for the independents is a hard one. When the local regulators see what has been happening, the companies will face pressure on their earnings, and to make ends meet they will have to reduce overhead costs at a time when they are especially in need of technical assistance.

It is a safe bet that the future for the smaller independents lies in consolidation with larger ones, or with the Baby Bells, which are no longer constrained by the provisions of the Kingsbury commitment.

Chalk up some more casualties to divestiture.

13

A National Disaster

Evil can be looked at as anything that causes or promotes disaster.

The Department of Defense recognized the evil in divestiture.

It knew the day was long past that our best defense was the Atlantic and Pacific. It knew the world had shrunk and the enemy could be on our shores within minutes instead of months. It knew our nation needed a short reaction time in this age of nuclear warfare. It was vital to have "White House to foxhole communications." Rapid information flow is essential and will continue to be essential.

While the federal government owns and maintains some communications facilities, the backbone of the system has been the commercial telephone network.

The government not only uses the same long distance circuits the private customer does, in addition it has circuits on the public network specifically designed for its use. These include the FTS (Federal Telecommunications System), AUTOVON (Automatic Voice Network), and AUTODIN (Automatic Digital Network). These and more than 95 percent of all the communications facilities used by the government—including for military command and control—depend on the same commercial resources used by all Americans.

Predivestiture, the national telecommunications network

was unified with built-in diversity, redundancy, alternate routing capability, and survivability. If a call couldn't get through one way, it got through another. Divestiture would change that with its LATA concept requiring interexchange carriers to interconnect only at the POPs, points of presence. It made the system more vulnerable, and that had the Defense Department worried.

No, the Department of Defense was not for divestiture. In the letter that Casper Weinberger sent to Attorney General William French Smith—the letter that got tossed into a safe waiting for Baxter's security clearance—the new secretary of defense wrote:

> "... The Department of Defense recommends very strongly that the Department of Justice not require or accept any divestiture that would have the effect of interfering with or disrupting any part of the existing communications facilities or network of the American Telephone and Telegraph Company that are essential to defense command and control."

The Defense Department didn't stop there. On April 8, 1981, Deputy Secretary of Defense Frank Carlucci wrote Baxter, by then cleared and on the job, "... it is the position of the Secretary of Defense that the pending suit against American Telephone and Telegraph Company be dismissed."

Baxter, as noted before, was not impressed and declared to the press that he intended to litigate the case to the fullest extent.

But this was too important an issue for the Defense Department to pack up its mess kit and disappear into the night. On June 30, 1981, it issued a report which said, in part:

> "DOD [Department of Defense] can unequivocally state that divestiture ... would cause substantial harm to national defense and security ... because it would substantially reduce, or eliminate entirely the incentives ... to engage in prior joint network planning and preparation to conduct centralized network management. ... We believe it would have a serious short-term effect, and a lethal long-term effect, since effective network planning would eventually become virtually nonexistent."

But the department's protests came to naught. AT&T caved in. Judge Greene, with no prior experience in telecommunications or national defense planning, took charge. Understandably this did little to assuage the Pentagon's fears. Army Colonel George H. Bolling, working in the office of the deputy under secretary of defense, was to express the government's continued concerns over the effects of increased network competition and divestiture in a paper entitled "AT&T— Aftermath of Antitrust." In it he stated:

> "Prior to divestiture, AT&T had acted as DOD's communications manager, engineer, integrator, controller, restorer and maintainer of telecommunications service. . . . AT&T was able and willing to act well beyond the normal contractor's role in providing priority government services. . . . The Bell System has fulfilled unusual and short-fused requirements that were actually failures within DOD to anticipate a need. . . . The divestiture has made the world's best network considerably weaker, less reliable, less responsive, and more vulnerable. . . . The decision to divest AT&T stunned DOD. . . . The Department of Justice had unmistakable messages from the Pentagon that proceeding with the settlement would jeopardize national defense and enlarge bureaucracy. . . . These arguments fell on deaf ears at Justice. . . . The lawyers from Justice, in their zeal to achieve a partial victory, were damaging the capabilities of their clients—the nation and the people."

What AT&T had given the government was end-to-end responsibility and control of telecommunications. The company took a problem and solved it. With shared responsibility, the buck gets passed. It's a sad fact but true. And AT&T was willing to go far beyond, as Bolling put it, "the normal contractor's role" because it was company policy, as can't be pointed out too many times, to put service before profit.

The unified Bell System was able to respond to the Department of Defense's needs. It could immediately issue the necessary directives and get all the component parts of any circuit or service together, even when that involved other suppliers or the independent telephone companies.

That's all changed now. It's not just AT&T anymore. There are multiple providers of intercity services and the long distance and exchange networks have been totally separated.

The government has lost the clout and resources of a single private sector organization that could deliver sustained telecommunications services—something the Defense Department called the "single manager" concept. The government can no longer go to AT&T and say "this is what we need" and know that it will get it. Now communications facilities acquisition and procurement requires detailed study before specifications can be drawn up and sent out for competitive bidding by the various contractors. This takes time, people—and more money. And, of course, when the government spends more money, the taxpayer ends up paying. And then, even when AT&T is the prime contractor for a particular service, it lacks the authority it once had to enforce its decisions. It must rely on persuasion to obtain cooperation from the subcontractors.

All this did not come as some sort of nasty surprise to the Defense Department after divestiture. In the period between the settlement agreement in January, 1982, and its implementation, the department fired off another memorandum to Justice on April 30, 1982, citing the potential calamitous effects of network fragmentation. These included the loss of end-to-end management and the loss of centrally directed engineering and development of technical standards. The Pentagon pointed out the additional costs in money and manpower the Defense Department would require to absorb the former AT&T workload. And again it stressed the expected degradation of national security.

But Justice was on a roll. Such minor issues as national defense and staggering increased costs to the taxpayer apparently didn't count for much and the divestiture went on.

So, one might ask, what's really the problem? Why couldn't the Defense Department step into AT&T's shoes and take over these coordinating responsibilities? First, it should be kept in mind that Bell was set up for coordinating efforts. Its vertical integration was perfect for this, and its employees were trained for it.

The Defense Department was starting, if not from ground zero, at least pretty close to it. It would have to recruit and train its own people—somehow lure them from the private sector. But while Ma Bell always paid top money and offered longterm, excellent career prospects, Uncle Sam was not in the same position. The government was going to need additional personnel in the area of technical specifications, procurement, and overall telecommunications management when it was already short of people with the needed skills. Furthermore, the skilled people it had already were being lured away to the private sector.

The Department of Defense knew it had a problem, but Justice didn't seem to care.

Judge Greene thought he could offset the negative effects of divestiture on national defense with the Single Point of Contact (SPOC) concept. The Baby Bells would have a centralized organization to meet the needs of national security and emergency preparedness. This was required.

He also recognized that there were some engineering, administrative, and other services that might be more effectively provided on a centralized basis. The Baby Bells, it was decided, would be allowed to set up such an organization, if they so chose.

The Bell System—this was during the interval between the agreement and divestiture, so there was still a Bell System— chose to combine the two functions into a separately incorporated entity that eventually was named Bell Communications Research, Inc., or Bellcore.

Initially Bellcore had an annual budget of a billion dollars and approximately eight thousand employees, of whom about 40 percent transferred from Bell Telephone Laboratories. Most of the rest came from AT&T staff organizations and a few from the local Bell operating companies. The work of Bellcore includes research and development for the Baby Bells, as well as providing them with staff and engineering support—very much as the AT&T staff organization did in the past. But there is one big difference. Previously the AT&T staff's suggestions carried with them the implied authority of the owner of the business, AT&T. That's no longer the case. Bellcore is owned by the Baby Bells, and they control what Bellcore will work on and make their own decisions on which Bellcore recommendations to adopt. Bellcore's functions are purely advisory. It retains no connection whatsoever with AT&T.

In terms of national security and emergency preparedness,

Bellcore's responsibilities were defined in part as:

- Coordinate development and implementation of uniform technical standards and nationwide emergency plans for the Bell operating companies.
- Expedite installation, testing, and restoration of local operating company services.
- Serve as a point of contact for other vendors to arrange for services provided jointly with the local operating companies.
- Operate continuously a national center to monitor the status of communications and to alert the Baby Bells during emergencies.

Thus, the SPOC's functions included twenty-four-hour surveillance and restoration of local operating company circuits on behalf of the Defense Department. But this still falls short of what AT&T used to do. The Single Point of Contact only has authority over circuits wholly or partially furnished by the Baby Bells. The circuits of the other companies such as MCI and AT&T and of the independent companies fall outside of Bellcore's domain.

Obviously, this was not acceptable. So another organization had to be set up in Washington. (Bellcore SPOC personnel have offices in Washington and at Bellcore's northern New Jersey headquarters.) The second group, called the National Coordination Center (NCC), has representatives from a number of communications firms, including Bellcore. Its purpose is to provide rapid contact between government agencies and most telecommunications carriers in the event of emergencies. While the NCC has already performed effectively in a number of situations, it lacks direct authority. It can't tell the various companies what to do. It has to talk them into doing it.

So what we have is two jerry-built organizations, costing the taxpayers and telephone customers a lot more money, and it's still not as good or as effective as Ma Bell's vunerable system.

To make matters even more troublesome, there is some question as to the future of Bellcore. Its existence depends on voluntary funding by the seven Baby Bells. In time, they will each inevitably become more independent in their thinking and will increase the size and competence of their own staffs. The different companies will develop their own priorities and the focus on uniformity will fade. When that happens, the companies will reduce their dependence on Bellcore. This is already happening at a fairly rapid rate. The key to Bellcore's survival as it is structured now will be its ability to provide expertise and services more economically than the Baby Bells can provide for themselves.

Colonel Bolling, for one, did not like the prospect of a declining Bellcore. He wrote in 1983, "If the authority and scope of Bellcore are reduced, standards will deteriorate rapidly, and lack of interoperability is the consequence. Interoperability is risked most in the private systems that bypass the national network. No policies mandate designing and engineering private telecommunications networks to meet standards beyond those essential for internal operations ... strengthening standardization and enforcing it are essential ... to compensate for the absence of a unifying Bell System."

The Baby Bells have the option of withdrawing completely from Bellcore with three years' notice. Nobody has pulled out yet, but that doesn't mean they haven't talked about it.

Should several Baby Bells decide to go it alone, Bellcore's function may be reduced only to the coordination of national defense and emergency procedures. Even then its effectiveness as a single point of contact is questionable since its authority and resources are derived from seven entities rather than the single source of the past.

As of January, 1988, several Bell operating companies were making plans to expand their own research facilities. U.S. West expects to open a $50 million research center in Boulder, Colorado, with a staff of fifteen hundred. NYNEX already has a research center in White Plains, New York, which it plans to expand considerably. Ameritech has also indicated it might build its own center as well.

Baby Bell research facilities are competitive and will keep their efforts secret from each other and from AT&T. There will be no wide open atmosphere as existed before divestiture in the unified Bell Laboratories.

And there is still another repercussion of divestiture about which the Defense Department must worry.

Hardened facilities.

Beginning in the 1950s, the Bell System voluntarily began a program to "harden" its facilities—that is, make them less vulnerable to enemy attack and natural disasters. Thousands of miles of coaxial cable routes were buried deep underground with reinforced concrete covers. Switching stations were built underground in buildings mounted on huge shock-absorbing springs, able to operate for extended periods without physical access to or from the outside world. They were designed to be invulnerable to anything short of a direct nuclear hit. The program also included many layers of alternate routing capability should a switching center or facility route be destroyed.

The government received no bill for the hardened facilities projects. Instead the cost was absorbed by all telephone users. The result was a highly diversified and protected communications system not only for the government, but for everyone.

That was then. This is now. The time for appeals to patriotism has passed. Competition is calling the shots. Neither AT&T nor any other carrier can be expected to pick up the tab for this kind of protection without directly billing the customer—in other words, the government. As a result, since the beginning of competition in the early 1970s, few new hardened sites have been built. Also with the establishment of the idiotic Point of Presence concept—that is, the interconnection between the Baby Bells and the inter-LATA carriers only at certain locations, whether that is the more efficient route or not—alternate routing capabilities have been greatly diminished. The arbitrary and totally ridiculous separation of long distance and exchange facilities—which makes no engineering sense at all—has cost the nation protection that had been built into the system over many years.

There is no escaping the truth.

Our country's communications network has suffered. It is more vulnerable and less secure than it was twenty years ago. The fragmentation of the system has, in the opinion of many military experts, seriously endangered the nation. Furthermore, future requests for emergency action, extra protection, or other specialized attention to government needs will be met with a bill.

There were probably quite a few smiling faces in the Kremlin the day Ma Bell died.

Actually, Uncle Sam is already getting estimates on the bill he'll have to foot.

Back when Ma Bell was alive and kicking, she gave the government a break on its telephone bill. Uncle Sam was billed at a lower rate than the average customer. But divestiture meant AT&T could no longer be so magnanimous. As a result the federal tab went up by $100 million a year, to almost $500 million.

The General Services Administration decided it was time to shop around to find a cheaper, high-tech telecommunications setup that would link the thousands of federal offices from the Virgin Islands to Hawaii. This would include more than one million phones, thousands of computer terminals, and an electronic call tracking and switching system.

Bidding has begun, and as this is being written, only one thing is fairly sure—whatever eventually goes into place, it won't be cheaper. The ten-year price range is now being estimated at from $20 billion to $30 billion, and perhaps even as high as $50 billion!

There is a lesson that can be learned here. Back in the glory days of Ma Bell, the government would have told AT&T what it needed. The company would have set about meeting those needs through a cooperative team effort. "Subcontractors" were part of the team. They didn't get a contract because they were the lowest bidder. No, the "subcontractors" were expected to keep cost in mind, but paramount to the effort was coming up with the best solution possible and carrying it out in the best manner. Quality was the primary goal.

But now telecommunications is in the throes of a wasteful laissez-faire, competitive cost approach. Now instead of the government telling Ma Bell what it needs, a tortuous, time-consuming, and incredibly expensive process must be followed.

First there is the preparation of enormously detailed

specifications covering minute details. This alone requires years of effort. Furthermore, these specifications are usually set in concrete. If a subcontractor happens to see problems with the specifications and wishes to change them, he'll usually refrain from doing so or be penalized by the contract's provisions. It happens all the time and it happened in the communications and control specifications for the San Francisco Bay Area Rapid Transit—BART—with calamitous results.

In this case, one of the subcontractors, the Union Switch and Signal Company, found serious flaws in the specifications but wasn't allowed to change them. So BART's communications and controls were installed as specified. The results? Collisions and deaths. Using a cooperative team effort in writing the specifications would have saved years of later investigation, reports, and modifications, and something in the neighborhood of $100 million.

Using a cooperative approach also does away with many of the problems of evaluation of prototype or product inspection. Because of the arm's length relations in competitive bidding, evaluation of a prototype can take years, and the inspection of even one part, should it be rejected, can cause numerous delays, returns, and modifications.

Then, after all this is done, the contract still has to be prepared. This voluminous document, on which a legion of lawyers have to nitpick away, is a formidable effort, and, of course, totally unnecessary in the cooperative team approach. The drawing up of this contract wastes time, effort, and, most certainly, money.

Of course, contract signing is not always the last step. Often those detailed specifications have become old and out of date and have to be changed. This requires a new batch of assignments and time to bring them up to date. So more time, effort, and money go down the drain without any beneficial side effect of improving or achieving economies as expertise develops in the progressing job.

Believe it or not, the merits of a cooperative team approach is no secret being revealed for the first time in these pages. It existed in the Bell System for one hundred years. And now it exists somewhere else.

Japan.

How Japan discovered or created the cooperative team approach is still unclear. But clearly it has used it to create the world's greatest industrial empire from scratch, providing superb quality products at costs below those of the United States, Germany, and the rest of the world.

Japan has come to realize some truths. Laissez-faire competition is obsolete, anachronistic, and destructive. And the introduction of laissez-faire competition in a utility operations is a return to the evils of the industrial revolution. On the other hand, the cooperative team approach promotes creativity, motivation, and the maximum participation of all workers.

But the United States clings stubbornly to its laissez-faire ways. And someday, after great expense and a ridiculous amount of time, the federal government will have its new telecommunications system. How super-duper it will be remains to be seen.

Our entire economy has been socked by divestiture.

There are those who predict that history will remember Ronald Reagan as the president who took the United States from being the number one lending country to the number one borrowing country. Divestiture, when the figures are studied, played its part in this tragedy.

Before divestiture, telecommunications was basically a Made in the USA operation. In 1981, the country imported about $7 billion in telecommunications equipment. Compare that with the $19.1 billion imported in 1987.

Trade in telecommunications products between the United States and Japan was out of balance in 1987 to the tune of $2 billion, with the gap increasing. Bruce A. Wooley of Stanford University has predicted that if the present trends continue, Japan will ultimately dominate the design and manufacture of telecommunications systems. It will then be Japan, not the United States, who will have world economic leadership in the information age.

Of course there are disasters other than military and financial. There are the disasters of waters that come raging down swollen rivers, the fury of wind swooping across a mid-western plain in a path of destruction, the shudders of the earth in a California valley—the disasters of a capricious Mother Nature. Ma Bell used to be able to deal with them.

Take June, 1972. Heavy rains in Pennsylvania and New York had streams and rivers far above normal levels, including the Susquehanna, which winds through central Pennsylvania on its way to Chesapeake Bay. In the neighborhood of Wilkes-Barre, the river takes a sharp turn. Here debris had accumulated, worsening the flooding effects.

The Susquehanna rose over its banks and the city of Wilkes-Barre was virtually cut off with water standing six feet deep in some of the main streets.

Being able to communicate with the rest of the world in an emergency such as this is essential. Unfortunately, there had been severe damage to two switching machines. One was in the central office for the nearby town of Plymouth. That office was completely out of service with deep water throughout the building. While this cut off service for some three thousand residents, damage to the second machine had even more serious repercussions. It was the long distance switcher in downtown Wilkes-Barre. With it out, communications into and out of the entire region were blocked.

Before water levels had even receded, the phone company established emergency manual facilities and restoration plans were underway. A team of Bell Telephone Laboratories experts examined ways to dry out the delicate crossbar switches in Wilkes-Barre without permanently damaging them—this particular situation had never happened before.

A mobile central office unit was borrowed from another state to replace the machine in Plymouth. Repair crews were brought in from a number of other Bell companies to fix the massive damage to cables and other plant. Western Electric shipped material necessary on a top priority basis.

Within a day, limited service was restored. Within two weeks, service was essentially back to normal.

The New York Telephone Company had its own disaster in February, 1975. Early one morning fire broke out in an eleven-story switching center building in lower Manhattan. Very quickly the combined effect of smoke, fumes, fire, and water knocked the entire building out of operation, resulting in more than 100,000 customers in that section of the city being totally without telephone service. On top of that, long distance service for much of the New York area was interrupted, and network management controls had to be applied. This meant a massive routing around the damaged facility to get calls through. In some cases calls from one side of the city to the other had to be routed to the West Coast and back.

A typical Bell System crisis mobilization followed. A dynamic New York Telephone Company vice-president, Lee Oberst, was put in charge of a task force and given carte blanche to deal with the emergency.

He immediately established an on-site crisis headquarters, and in less than twenty-four hours, emergency service to hospitals, fire and police stations was restored. By day two, an army of four thousand workers had been mustered from throughout the Bell System to work on restoral of service in twelve-hour shifts.

At the same time Western Electric began manufacturing or redirecting from other applications huge quantities of replacement switching equipment and shipping it directly to downtown New York. Bell Laboratories personnel searched for faster, better ways to make repairs.

In short, the full capabilities of the unified Bell System were called up—and it delivered a miracle. Full service was restored within three weeks.

It took a remarkable mobilization of talent, centralized direction, and interchangeability of equipment and procedures, but the near-impossible was done.

That was the way it was in the 1970s. But what of the 1980s and beyond? Local disasters, whether natural or man-made, often require efforts beyond local capabilities. The need to shift and redirect telecommunications resources to meet these emergencies is vital. The Bell System is fragmented now. Bell Atlantic or Pacific Telesis or NYNEX have smaller pools of resources. And none of the Baby Bells can expect the response of the old Western Electric in getting replacement parts to the scene as needed. All this adds up to a diminished restoration capability. And no matter how one looks at it— that's a disaster.

Ma Bell? She's out of here, gone, history. Shed a tear and move on, look to the future and ask the question—will the country be able to cope with a true national emergency considering the structural changes in the telecommunications system?

At this point, we can only cross our fingers, hold our breath—and hope for the right answer.

14

The Customer's Dilemma

True Story Number One.

Eric built a small workshop/office in a shed in his backyard. He wanted to do a little puttering, out of the mainstream of his family's traffic, but not out of touch completely. What he needed was a telephone installed in the office with an intercom to the main house.

It seemed a simple enough solution.

With that decided, Eric called his local Bell business office and described what he wanted. The Bell representative, while very pleasant, couldn't be much help. The local Bell company, the representative explained, was no longer in the business of manufacturing and selling telephones, and, no, she couldn't tell Eric where he might find a good telephone/intercom set. (Bell employees are scrupulously careful not to refer customers to AT&T or any other supplier, to avoid being accused of favoritism.)

Well, okay, so Eric set out on his own in search of a set. He found what he was looking for at the AT&T phone center. Wonderful, he thought, he was home free. Just arrange for AT&T to put in the necessary wiring and jacks.

But there was one small problem. AT&T is not in the business of running wires from his house to the shed and installing the proper jacks and mounting boards. No, that is in the jurisdiction of the local Bell company.

Another call. Yes, the local Bell company was prepared to run the wires—which would set Eric back more than the cost of the phone—but, alas, it wouldn't install the conduit in which the wire had to run. No, an electrician was needed for that—unless Eric wished to do it himself.

Since Eric definitely did not wish to do it himself, he hired an electrician. Within a mere week, conduit and mounting boards were installed.

But, of course, the wires and jacks still weren't in place. Another call to the local Bell company. Another few days. And finally the wires were run and the jacks mounted.

Eric was sure he was almost at the end of the tunnel. Back to the AT&T store to pick up the set, back to the shed, to plug it in and—oops! The phone plug didn't match the jack.

Still another call to Bell. More days pass. And then, at long, long last with not a small degree of frustration, the installation was completed.

Eric's final score: three weeks of waiting, calls to Bell, calls to the electrician, visits to the AT&T phone center—*and* more than $300 out of pocket.

Eric's score before deregulation?—one telephone call, two or three days waiting—*and* less than $50.

Eric isn't the only telephone customer to be put through the wringer since deregulation. In one way or another, we all have. In the old days, the customer needed two numbers to get what he wanted—one to the business office for installation work or any billing problem, the second to repair service for anything that might have gone wrong.

Now there is a plethora of choices and decisions:

About the telephone instrument—

Should the customer buy his set?

If so, from whom?

Besides price, what factors should he consider? How does he judge quality?

Should he buy a maintenance plan? What should it cover?

Should he chuck it all and rent? If so, who should he rent from?

About inside wiring—

Should he install it himself?

Should he hire a contractor?

Should he gamble that he can maintain it himself and be prepared to pay Bell in the neighborhood of $90 an hour for repairs if he can't?

Should he take out insurance on the wiring in the form of the Bell service contract which runs anywhere from $6 to $40 a year?

About long distance carriers—

Should he designate a primary carrier?

If so, which one?

When should he use a carrier other than his primary one? To do this, what prefix codes should he dial?

When should he change his primary carrier?

How does he do that?

About repairs—

Does he take his phone back to where he bought it?

Does he call the local Bell repair service?

Does he attempt to fix it himself?

Does he call Judge Greene for advice and counsel?

About his bill—

Should he carefully analyze the many local and long distance components of the bill with the various surcharges and what-nots?

Should he just accept it at face value, send in the check, and hope it gets there before the local Bell has churned out the automatic suspension notice?

Should he pay $200 an hour and have an accountant look it over?

Deregulation and competition have thrust the customer into the telephone business, whether he wants to be there or not. It takes more time, more money, more aggravation, and when it's all done, there's no guarantee—or even much likelihood—that the finished product and service is going to be as good as what the old Bell System provided. In the area of terminal equipment, deregulation and competition and the way it was introduced have degraded service and raised cost.

The time—in the past, when Ma Bell lived.

The place—a toll maintenance center.

The characters—testboard operators, the men who sat perched on high stools in front of an eleven-foot tall array of jacks, switches, lights and meters. (Later, when computer consoles took the place of the mechanical devices, women were added to the cast as well.) The operators' job was the end-to-end testing of intercity circuits and equipment.

Back in these simpler times, all equipment, whether in the intercity network or on the customer's premises, was owned by the Bell System. There was, therefore, no question or ambiguity over who was responsible for keeping the equipment running properly. Bell maintained it. Period. It was up to the testboard operators to isolate and identify problems. In doing so, they communicated with other testboards and with equipment operators all across the country.

At some indefinite time in the past, some anonymous testboard operator made history. After determining that the problem he was working on was actually in someone else's jurisdiction, he announced to the person down the line, "It's all yours. The trouble's leaving here okay!" The phrase, odd though it was, caught on, and became a greeting among the fraternity.

Before divestiture, when the tester made his announcement, he was confident that someone would pick up the problem wherever it existed. Today, as a result of divestiture, the network is chopped up and ownership divided, customers own the terminal equipment, and responsibility is fragmented among a number of organizations.

Now when the tester announces "the trouble's leaving here okay," he can no longer be sure who, if anyone, will respond and accept responsibility.

Much of the blame for this sorry state of affairs can be plopped right on the doorstep of the FCC, which did not seem to be operating with any coherent plan or goal, beyond "let the competition begin."

It went by dribs and drabs toward introducing competition, starting with its Carterfone decision in 1968, which was the first significant break in the long tradition of Bell's end-to-end responsibility, so typified by the testboard operators.

To many the Carterfone decision might have seemed innocuous enough. AT&T was ordered to revise its tariffs to allow this device to connect private two-way radio systems with the telephone network.

When it did, the new tariffs called for using protective interface devices (or couplers), to be provided by Bell, when such interconnections were made. AT&T hoped in this way to insure the integrity of the whole system.

But Pandora's box had been opened, and during the next ten years, successive FCC decisions were to eventually deregulate all terminal equipment and permit interconnection without the protective devices. In truth, the FCC had no business concerning itself with terminal equipment. It had been chartered to regulate interstate traffic. The commission clearly violated that charter by invading every household and business establishment in the country with its various decisions, though that never seemed to bother it.

The term terminal equipment—aka station equipment and customer premises equipment (CPE)—includes more than just the common telephone instrument. Under this heading fall private branch exchanges (PBXs), key telephone systems, facsimile machines, data modems, automatic dialers, answering machines, and any other equipment on a customer's premises that connects to the telephone network in one direction and interacts with the user in the other. This terminal equipment represented roughly $60 billion, or about 25 percent of total pre-deregulation telephone investment (Bell and independent). The current figure is well over $100 billion.

Not unexpectedly, after deregulation many companies rushed to get a piece of the terminal equipment action. At one point, some two hundred manufacturers were making telephone sets. That was winnowed down to fewer than fifty, with AT&T retaining a little more than 20 percent of this $1.25 billion market.

One step up in complexity from the standard telephone is the key system. This lets users select any of several telephone lines by pressing buttons or keys at the base of the phone.

This is currently a highly competitive $2.5 billion market, with AT&T holding a 30 percent share. Three of the top six companies in the field are Japanese.

When a key system is not big enough to meet a customer's requirements, the next step up is the PBX. There are approximately thirty-five companies who divide up this $4 billion market, with seven accounting for 90 percent of it. Of these, four are foreign. AT&T has about a 25 percent share.

The costs of PBXs and key equipment is now two to five times as much, although some of that rise is due to increased sophistication of the machinery.

So, clearly, competition was a windfall—for foreign companies. Before deregulation, telephone equipment was Made in America. Now it can be made anywhere—and often is. Even AT&T is producing some of its telephone sets in China.

Which brings us back to the old question.
Why?
Why did the FCC decide that competition in the telecommunications field was good for the country's industry? In search for an answer we must return to the Toaster Theory.

It would seem that the FCC thought competition was good for the industry because it wasn't bad for the industry. That is, what was the difference if inferior equipment was plugged into the system? What harm could there be?

It failed to understand that the total telephone network is a single vast computer distributed all across the country. Having different parts of this computer designed to different standards by any number of manufacturers is sheer lunacy. It invites chaos. An automobile is a far less complex "system" than our national telecommunications network. Yet who in his right mind would choose to drive a car the parts of which were designed independently by different companies, and then thrown together without overall coordination of how the parts interacted? Only someone with strong suicidal tendencies.

As has been pointed out, despite popular opinion, Western Electric did not have a monopoly on terminal equipment before deregulation. There were other suppliers as well. While

it's true that Western Electric made almost all the home telephones used by the Bell System, a good deal of other terminal
equipment was manufactured by other companies—but in
conformity with Bell specifications. Even with common telephone instruments, many components were made by others
and assembled by Western Electric. Also, for many years non-
Bell equipment was used on telegraph, data, radio, and television services. The key factor, then, was not whether Bell manufactured the equipment; it was whether all the components
of the network conformed to established system standards.

The Toaster Theory ignored that fundamental necessity,
and in doing so, created six problem areas—

1) physical damage to the network and its users
2) adverse effects on service to others
3) division of responsibility for service
4) stymied technological progress
5) confiscation of capital
6) diminished response in emergencies.

The potential physical damage that interconnection can
cause has received the most attention, yet it is probably the
least significant problem, at least for ordinary residence installations. And the attention given it has diverted both the
regulators and public from more serious issues.

The way some would have had it believed, every time a customer plugs in a non-Bell phone, he's at risk of being injured
or even killed. Obviously, it is possible to damage the
network—wires, cables, central office equipment—by connecting a device that sends too much current and voltage
into the system. And it's also obvious that the excess voltage
could hurt an unfortunate someone who comes in contact
with it. The potential for these calamities is only increased
when there are no standards applied to the equipment connected into the network.

But when push comes to shove, it's very unlikely this will
ever happen to a residence customer. Unfortunately, when
Bell was pleading its case before the FCC, Bell witnesses tried
to document such occurrences—without much success. In
the end what suffered most was Bell's credibility.

The threat of physical damage becomes much more real,

however, when a large business or military customer with extensive wiring on his premises is considered. Often the customer does his own wiring and may not adequately separate the telephone and high voltage facilities. In many cases the same pole lines or underground conduits are shared. In these situations there are very real, potentially deadly, dangers.

It is also imperative that detailed and accurate records be kept showing how equipment components are connected and which cable pairs are used. This sort of meticulous documentation has, historically, been done very poorly at many military bases. This was partially because changes in cables and equipment were being made constantly and partially because poorly trained and inadequately supervised personnel were doing the work.

AT&T conducted a nationwide study of customer premises equipment at military bases in the 1960s. At many locations it was found that the records were so bad and the cables so poorly spliced that the simplest repair jobs were major undertakings. Finally, to restore adequate service, the entire cable plant at numerous locations had to be replaced. It was a clear example of how customer ownership could become a veritable rat's nest.

In its Carterfone decision, the FCC undoubtedly thought it had gotten around the problem of interconnection's potential adverse service effects by requiring the protective interface devices. The device would prevent undesired or excessive signals from entering the network.

And they might have effectively done so if AT&T hadn't made a serious, almost arrogant, blunder that would eventually contribute to the downfall of Ma Bell.

The only place competitors could get the protective interface or coupler was from AT&T. So what did AT&T do?— whether deliberately or not didn't matter in the long run—it delayed making the devices available, effectively hamstringing the competition. Without the coupler, the competing companies could not legally sell their equipment. This led to challenges as to whether the interface devices were even needed in the first place.

In 1974, the FCC decided this issue. The protective device

would no longer be required. Instead it instituted a program of registration or certification of equipment. Thus the terminal equipment field was opened to any who wished to enter with the caveat that the equipment must be manufactured to prevent "harm to the network." This included physical harm as well as harm to the service of others and was defined in electrical terms—impedance balance and transmitted power.

What was omitted was any mention of quality of performance. Technically by the FCC standards a pair of tin cans tied to the network with a string would qualify. On one hand, bad equipment might not blast someone into their grave. On the other, any defect in a telephone instrument is a defect in the entire network. And that's bad for everyone else's service, even if that only means that others can't get through to the person with the faulty phone, wasting the caller's and the network's time. Good quality and well-maintained telephone terminals are in the best interest of everyone on the network.

None of this is to say that all non-Western Electric equipment is inferior. At one time that was pretty much true, but not any longer. In recent years a number of suppliers have marketed equipment as good as Western's and, in some cases, with more advanced features. At the other end of the spectrum is the Cheapo Charlie junk. Connect that into the network and there are bound to be problems.

Take the increasingly popular portable telephones. High frequency radio signals, capable of transmitting from 200 to 500 feet, are used in place of wires to connect the handset to the base unit. So, a sunbather doesn't have to leave poolside to answer the phone. Nice. Nice if properly designed. If not, it's a mess. Then it is not only hard to carry on a conversation, but the radio transmissions of these cheap models may interfere with other telephones, with television sets, and with other appliances. What's more, good or bad, portable telephone users must be discreet. Conversations are not always private. So if it's something the next door neighbor shouldn't hear, it shouldn't be said.

Then there are those privately owned coin telephones that are popping up all over the place. Today anyone can make, sell, and install these little wonders. Owners of these telephones pay the telephone company a monthly charge and then keep the coin revenues. Sounds good so far. But wait. These phones operate independently of the network. That means they must have within them provision for generating dial tone, recording the coins dropped in, and computing long distance charges based on the number dialed and where the telephone is located. At the end of the month, the owner of these coin telephones gets a bill from the local Bell company and from his long distance carrier. He has to make sure that whoever used the phone during the month put in enough money to at least cover the costs.

All this necessitates a complex and costly instrument, since all these features can be and are provided more economically on a centralized basis with Bell-owned coin telephones. So right away, the cost of a call from one of these privately owned coin phones is usually higher. Beyond that, they're less reliable, partly because of their complexity and also because they're dependent on local electric power. As a result, service suffers. There's also a greater chance of the machine being on the blink, eating coins when a call doesn't go through, and being an all-around nuisance to use. And one mustn't try calling information. The call costs the same as any other call, and often the quality of transmission is so poor it's impossible to hear what the directory assistance computer is saying. Which means a second call—and a second charge.

There is yet another downside to poorly designed, installed, or maintained equipment, an aspect that has been generally overlooked—and totally ignored by the FCC. Say the telephone company installs a PBX for a business customer. This PBX has been carefully engineered to provide a specified grade of service for that customer, and also to prevent backups and overloads in the public network. But say the customer decides to take over this design responsibility— generally with little or no telephone traffic engineering background but with the directive to keep costs down; the result is often poor service for him and degraded service for others.

In 1976 John deButts took this concern of "harm to the network" before a Congressional committee. He told of a PBX, owned and installed by Citibank in New York City, that had been badly engineered. When the PBX went into service, the rest of the city's service was almost shut down from the ensuing telephone traffic jam.

True Story Number Two.

A member of Fred's family had to reach him, urgently. The person tried calling. No answer. No answer for hours. The caller was sure Fred was home and became concerned— concerned enough to get into his car and drive an hour and a half to Fred's house. There he found an unhurt, healthy Fred. What had happened? Hadn't he heard the phone ringing?

No, Fred hadn't because both of his phones had been unplugged.

It was all kind of silly, in a way. Fred had a large house with modular phone jacks in every room, but only two phone instruments. Fred and his wife would move the phones from room to room, depending on where they wanted to make a call. On this particular day, each had started to move a phone, were distracted, and didn't plug either back in, leaving them completely cut off from the telephone network and from the rest of their family.

Silly in a way, but serious as well. Fred's relative had to get in touch with him. It wasn't frivolous. It was important. Back in the old days, when Ma Bell owned all terminal equipment, this situation could not have happened. Then Ma Bell required the customer to have at least one telephone or ringer permanently wired. He could have as many jacks or portable phones as desired, but somewhere a ring had to be heard.

Back then Fred's service was protected and the entire network was better for Fred being a permanent part of it. That was back then. Here and now, Fred has the luxury of designing his own system—and forgetting to plug in his phones.

True Story Number Three.

Mrs. A., on the other hand, has no luxury at all. Eighty-eight, she lives alone in a small home on Pittsburgh's North Side. She has no living relatives and depends on daily visits from Meals on Wheels for food and on her telephone for emergency communication in case of illness or accident.

One day Mrs. A. was unable to get a dial tone. She asked a neighbor to call the local Bell company's repair service. Repair service found that Mrs. A.'s line tested OK and said she should take her telephone instrument to AT&T for repair. That was easier said than done because her telephone was hard-wired and she wasn't about to cut the cord herself. Finally, another neighbor cut the cord for her, but for two weeks Mrs. A. was without the use of her phone. In the old days, one call and a few hours would have had her back in service. Divestiture has been hard on the elderly, the poor, and the disabled.

Responsibility. Give everyone responsibility and no one is responsible for anything. Telephone customers are learning this the hard way. For one hundred years, the Bell System assumed end-to-end responsibility for telephone service. End-to-end responsibility was not some Madison Avenue campaign slogan. It was reality. Have a problem, call 611 and the repair man was there pronto—free of charge.

Which number does the customer call now?

True Story Number Four.

Ray has run a consulting firm from his home for years. And for years he was quite happy with his key telephone system with its two lines and four telephones. All the equipment had the Western Electric imprimatur, and had been leased from Bell of Pennsylvania. The service had been excellent.

After deregulation, Ray was forced to purchase the equipment for approximately $3,000. So far, except for the expense, no problem.

At about the same time, the local Bell central office equipment was modernized—a thirty-year-old #5 Crossbar switcher was replaced by the #5E time division electronic switcher—AT&T's most up-to-date product.

It was then the problems began. Ray was in the habit of calling a customer who lived in another state, some one hundred miles away. He began to consistently reach wrong numbers when calling this particular customer—and almost always the same wrong number. Ray suspected the new switcher might be the source of the problem, and he reported it to the local repair service.

Repair service, presumably after some testing, reported

that there was no central office problem and suggested the telephone instruments be checked. After a bit of research to find the proper number, Ray called AT&T, which dutifully sent out a serviceman. After several hours checking the equipment, the AT&T man reported nothing wrong with the instruments.

So what was it? The repairman suggested the glitch might be in the AT&T long distance switcher. Ray should report his problem to AT&T Communications. The repairman couldn't do it since AT&T Communications was separate from AT&T's equipment subsidiary and kept at arm's length.

Ray made still another call, was transferred several times, put on hold (with its accompanying charming music) more than once before finally getting through to the right party. Having done all this, Ray thought it prudent to again complain to Bell of Pennsylvania.

This all took place over several months. Then one day, the problem disappeared. Ray would dial his customer and get his customer. Who fixed the problem and how will forever remain a mystery to Ray. All he does know was the entire operation cost him a great deal of time and annoyance.

In one respect Ray was lucky. There was no out-of-pocket expense involved. For others a wrong decision means time and money. If the local telephone company repairman comes to the house and discovers the problem is in the customer-owned telephone, the customer has to pay for the service call and still has a broken phone. To get the instrument repaired takes two trips to the service center.

Time and money are squandered trying to pin down responsibility, and still more getting the actual repairs done. Even calling for repairs is complicated. Instead of the old 611, the customer now must decide between several different seven- or ten-digit numbers. Can this really be called progress?

Many large companies, with properly trained staffs, are in the telephone business by choice. But pity the millions of small businesses and residence customers who lack the knowledge and desire to make the myriad telecommunications decisions they are now forced to make.

Some may applaud what they deem a new freedom of choice, but surveys indicate most people are not happy with it and do not understand it. Many polled feel they are being denied another freedom—the freedom not to have to make all these choices.

Their monthly bills have become not only larger but also more complicated. An advanced degree in accounting is needed to understand them. Generally AT&T pays the Baby Bells to do its billing, and the charges for the separate companies' services are listed on separate attachments. Added to that are listings for telephone rentals, local central office connection, inside wiring, and individual local calls as the trend for each service to pay its own way continues.

As if all that wasn't enough to cram on the bill, there can also be the incentive options such as AT&T's Reach Out America program. Pity the poor postman who has to deliver these tomes.

AT&T is planning to do its own billing to eliminate having to pay the Baby Bells for this service. While this will simplify the bill, it will double the number of bills the customer gets and the number of payments he has to make.

If they don't already know, customers are now sensing that the fragmentation of responsibility has had a negative impact on service. When all telephones on the Bell network were made by Western Electric, on the average seven million repair calls were logged per month. Now with junk telephones flooding the market and making their way into homes and offices, the trouble rate can only go in one direction.

As far back as 1973, John deButts in testimony before the FCC was able to cite a study done on more than 800,000 private lines. He said, "Current studies indicate that intercity private line links with at least one customer-provided terminal generate trouble reports at a rate at least 50 percent higher than do serving links equipped with Bell terminals only."

More than likely, there will be a lot of trouble never reported on these $9.98 pieces of junk that are made with no quality control. They'll be discarded like a Kleenex. So that great bargain, in the long run, is going to cost the consumer a lot more when it has to be repeatedly replaced.

Of course, the residence customer is only dealing with a few self-owned instruments. Pity the big metropolitan businesses. In the past the local telephone company often had maintenance rooms on site at major customer locations with a permanently assigned staff of installers and repairmen. Big Company had a telephone problem on the 17th floor, and up the elevator went the telephone repairman to fix it. Travel time was in seconds or minutes, at most.

Today, the business customer first has to figure out which of the many equipment suppliers he should call for repairs (and hope he guesses the right one). No single one is likely to have a sufficient concentration of equipment in one building to warrant an on-site maintenance force.

So having placed the call, the customer has to wait for the repair people to get to him—delaying correction of the problem and adding the cost of travel time, motor vehicles, and related overhead.

With a conservative estimate of 20 million telephones in urban locations of this sort, each needing repair or rearrangement once every two years, with a mile of travel time from the maintenance center, the grand total is some twenty million miles of travel and more than 500 man-years of wasted travel time annually. This is like having a half dozen mailmen serving the same block.

On top of all this, the businessman must worry about the privacy of his communications and his overall security when so many people from so many different sources have access to his equipment and office.

Progress?

"Looking ahead, I am concerned that ... fragmentation ... would represent an obstacle to engineering the 'intelligent network' of tomorrow. How 'intelligence' will be distributed among the terminals, switching nodes, and transmission paths of the network we can't currently predict. It would be regrettable if arbitrary corporate boundaries precluded our doing so in an optimum way."
—CHARLES L. BROWN, AT&T Chairman, 1979

In many other industries, competition does stimulate technological innovation. General Motors, Ford, and Chrysler continually try to out-innovate each other and in the process come up with new features.

And in one respect—variety of new products available—it has done that in the telecommunications industry. But for the most part, the introduction of competition can only stymie technological progress. Once again it's a question of the nature of the beast.

Because the telecommunications network is such an interrelated and interactive system, innovation in one part of it must take into account the rest of it. There are tradeoffs in cost and performance between terminal equipment, loop plant, switching equipment, and intercity transmission systems. Once responsibility gets divided in these different parts, it becomes very difficult to optimize technical progress in the whole.

Just take that least sophisticated piece of equipment—the home telephone. When Ma Bell owned it, she fixed or replaced it free of charge. And when technology advanced, outdating the old model, it was Ma Bell's prerogative to replace it with a newer and better instrument. Which over the years she did. In fact, between 1920 and 1960, three complete replacements were made—to every one in the system.

Consider the situation today. The customer has plunked down his money for the telephone set. Should a better, more advanced model come along, it's up to him to go out and buy it. But his attitude most likely would be, why? The old one works adequately, so why pay for another one? The only thing that might prod him into getting a new phone is appearance. In short, the home telephone instrument is locked into today's technology.

The consumer and the country can no longer expect the overall systemwide benefits reaped from the planned and orderly introduction of newer equipment. Service is forever tied to the performance of the telephone set the customer buys. And this is surely a barrier to progress.

That is on the simplest level. On the larger, more complicated scale? There is no longer a Ma Bell with her tremendous size and financial resources to pursue long-term and costly research which smaller organizations could not afford. Now, with competition increasing the uncertainty of the market, short-term profit and market advantage become the driving factors. It will now be "what's in it for us?" rather than "what's in it for the network, for the customer, for the country?" Research, development, and ultimately the innovation process are bound to suffer.

It almost seems that, like in a silly comic strip, a lightbulb popped on in the collective head of the FCC. The grand idea was "Competition—good." And that was that. The lightbulb was turned off and no further thought was given to the concept. Once accepted that competition was desirable in the telecommunications industry, whatever came down the road was going to be good as well.

That's not real life. In reality, one thing leads to another to another to another. Reasonable men and women look beyond the immediate result to long-term ramifications. They explore impacts and nuances. They make allowances.

Using its misguided economic theories, the FCC said let there be competition in the terminal equipment market. And so there was competition and a whole new industry was created. That looks fine on paper. It looks good as far as it goes. But in reality what happened was that much of that whole new industry went overseas at the expense of American jobs with resulting massive layoffs at AT&T. Our foreign trade balance has taken a beating. And it didn't take a crystal ball to foresee this, but somehow the FCC couldn't see beyond its concept.

There was another consequence of customer-owned equipment for which the FCC made little allowance. When customers were allowed to substitute their own equipment, what became of Bell's? It had to be removed, and much of it junked. But the FCC did little to shorten its allowable depreciation intervals. When there is a short-term depreciation schedule, a company takes big write-offs up front. But Bell had been working on a long-term depreciation schedule. This meant the equipment was scrapped before being fully written off. And the FCC did not compensate Bell for that.

In essence, AT&T was the victim of capital confiscation.

Furthermore, in the PBX field, AT&T's competitors were able to use Bell-developed technology to leapfrog ahead of what AT&T had to offer. The competing equipment had more advanced features because it was using more up-to-date electronic equipment. Meanwhile, AT&T had perfectly serviceable electromechanical PBXs in place it didn't want to junk until they had been fully depreciated. The government was actually promoting unfair competition and, in effect, again confiscated Bell System capital.

The immediate victim of this confiscation was Bell. However, eventually the costs flow to the consumers. Once again, a government decision has benefitted no one except a few manufacturers, many of them overseas, and the rest of us have had to pay the price.

The Internal Revenue Service gave businesses another incentive to abandon PBXs they were renting from Bell in favor of buying their own switching machines. Through investment tax credits and accelerated depreciation, purchasers of new PBXs could write off 30 percent of their cost in the first year alone. With the advanced features being offered by the other manufacturers, many businesses abandoned Bell altogether.

But the average business customer is not knowledgeable enough to make proper cost evaluations between Bell-provided service and owning his own PBX equipment. And equipment salespeople are hardly the ones to ask. Some hidden costs include:

- expense of supervising the operation and maintenance of the system
- cost of expanding the capacity of the system
- cost of telephone movement and changes
- cost of modernization and future additions to the system.

True Story Number Five.

A PBX salesman got the management of a Philadelphia hospital to believe it would save a lot of money by buying equipment from his company rather than continuing to rent from the local Bell companies.

It's not unheard of, and was the case here, that a salesman will cut corners to keep the price down, in hopes of making the sale. The system he sold the hospital was undersized and did not have enough call-handling capacity.

That led to real headaches that two aspirins couldn't clear

up. During heavy calling times, only half of the incoming calls were getting through. The rest got busy signals even though the telephones they were trying to reach were idle. The problem was more apparent to the general public than to the hospital staff, but the public soon rectified that with many complaints.

The PBX vendor was called in to redesign the equipment. There wasn't much he could do because the basic limitations of the machine made it impossible to increase the call-handling capacity to the required level. After struggling with the problem for a few years—and probably with a lot of other similar ones with other customers—the PBX company went out of business.

The hospital junked the PBX it had bought and went back to Bell. The hospital paid for its mistake in dollars and cents. One has to wonder what the bill was for all those people with emergencies who couldn't get through to the hospital.

True Story Number Six.

PBX vendors found green pastures in motels and motor lodges. They were dealing with people who had little knowledge of telecommunications beyond dropping a quarter into a pay phone. The salesman usually could make a convincing economic case for buying his product. Add to that the attractive service features their PBXs offered (but not necessarily delivered) and the sales were easy.

It was the outcome that was hard.

The owner of a 100-room motor lodge had bought such equipment consisting of a Japanese-built PBX along with terminal equipment from four different manufacturers. The package was engineered and sold by a small telecommunications firm and was supposed to include conferencing equipment, message recording and delivery, and detailed call-charging information.

The PBX went in but calls didn't. After cutover the equipment failed completely on several occasions allowing no calls to get in at all. Some of the failures lasted a few hours, some a day or more. There were also times when most outgoing calls were not completed.

Finally in desperation, the motel owner hired a consultant.

After extensive analysis and testing, the consultant came to these conclusions:

1) Approximately 20,000 incoming calls had been lost in the PBX's twenty-one months of "operation."
2) Approximately 13,000 originating calls failed during that period.
3) There were more than 200 other specific failures of the equipment.
4) The basic problem was that the various parts of the system weren't properly integrated.
5) The owner should replace the entire installation and seek damages from the supplier.

In the case of PBXs, the IRS and the FCC didn't make life easy for Bell, but AT&T may have been its own worst enemy.

In the early 1960s, the company introduced a new service for business customers called Centrex. It provided many of the advantages of a PBX system without the drawbacks. Centrex uses a portion of the Bell central office for the customer's in-house calling. With a PBX, a switcher has to be installed on the customer's premises. With Centrex, the concentration effect of the PBX is lost since each of the customer's telephones has to be connected to the central office by an individual pair of wires. This means Centrex service is only practical if the customer is within a mile or two of the central office. But this was not a serious limitation in urban and industrial park locations. And in some cases where there was heavy demand, a new central office was established primarily to serve specific Centrex customers.

At first Centrex was popular only among large businesses. But gradually even customers with only a few dozen lines began using it because of its advantages. For one thing, a customer didn't have to give up the space on his premises a PBX took. He had the benefit of the central office's full emergency back-up power and twenty-four-hour on-site maintenance. And should his company move, he didn't have to deal with the cost and complexity of relocating a PBX.

Then, too, his system wasn't static. If the central office equipment was modernized, so were his service features—

automatically, with no cost or action on his part. Maintenance and new service was faster and cheaper since the central office switching location was continuously manned or under electronic surveillance.

Centrex was good for Bell, as well. Once a customer was using that system, he wasn't as likely to replace it with a PBX, which might be an AT&T PBX or it might be a competitor's. Even if a customer did switch to a PBX, it was more likely that Bell's equipment investment, because it was located in its central office, could be reused.

Given all this, it seemed logical and preferable for Bell to promote and improve its Centrex system. After all, Bell had the central office locations without competition.

But no, this is where AT&T shot itself in the foot. In the mid-seventies, some so-called marketing experts were brought in from other companies—notably IBM—and they decided that Centrex should be downplayed. AT&T should concentrate on the PBX field, and go eyeball to eyeball with the competition.

So Centrex took on the role of Cinderella and her stepsister, the PBX, got all the attention. Bell scientists and engineers began concentrating on stand-alone PBX technology at the expense of Centrex's.

Bell salespeople went out and actually unsold Centrex. Bell engineers saw the folly of the new policy, but were unable to get it changed. In retrospect one almost has to wonder who those "experts" were working for.

Take what happened in Danville, a town in rural north central Pennsylvania. It had a #5 Crossbar central office switcher which served some four thousand residences and small business customers, along with approximately two thousand lines of Centrex service for a large medical center.

Centrex was in place and working well, but the Bell marketing people decided the medical center was a prime prospect for Bell's Dimension PBX. So the salesman—who was in line for a nice commission if he could sell the Dimension—descended on the medical center, proclaiming that Centrex service was obsolete and PBX was the way to go.

After mulling it over, the medical center people agreed with him—abandoning Centrex and going with a PBX from a competing supplier. The unhappy result for Bell was not only did it not sell a PBX, but it also lost the Centrex revenue. A significant portion of the Danville central office capacity plus several miles of outside plant cable were idled—without much chance of being used again.

This may sound like the proverbial broken record—but in the end the ratepayers of Pennsylvania were saddled with the carrying costs of this now surplus and unusable equipment.

This was hardly an isolated case. It happened time after time, and in large part blame here goes to Bell's management.

It wasn't until after 1980, when many Centrex installations had been disconnected and the opportunity for new ones missed, that the Bell System finally realized what was happening and began to reemphasize Centrex. By then there was a Centrex technology gap, so research and development had to be accelerated. By 1987, however, Centrex had made a considerable comeback and was being marketed aggressively with TV and print ad campaigns.

There are still two stumbling blocks for Centrex. One is the FCC-mandated access line charge for each individual line connected to a central office. Because the Centrex lines all feed into the central office while the PBX does not, the access line charge for a Centrex is higher.

The second thing that detracts from a Centrex's desirability is that the local Bells have not been allowed to offer enhanced services—message recording and forwarding, database operations, etc.—except through separate subsidiaries. These services can be provided on a stand-alone PBX, but not legally on Centrex, where all the functions are combined in a single machine.

The local Bells have been trying to get a modification of both the access line charges and the enhanced services restrictions. If they succeed, then watch out PBX, Cinderella Centrex will be wearing the glass slipper and moving into the palace.

And finally, the last point, where the Toaster Theory fails to stand up to scrutiny. It's in those times of emergencies, when everything is not working just as it should, when Mother Nature is not cooperating, or the electric company's generator decides to take a vacation.

Ma Bell had recognized that things don't always go as planned—and if you want something done right, you do it yourself. So she insisted that the telephone network operate independently. Telephone company installations—central offices, repeater stations, microwave relays, etc.—invariably have enormous battery plants to back up the commercially supplied power. But it didn't stop there. The backup battery plants had backup gasoline or diesel-powered engine-generator sets with several weeks of fuel supply on site.

Terminal equipment was also designed to stay in service without an external electric power source. Ordinary telephones were powered over cable pairs from central offices. Key systems and PBXs were powered the same way or by local batteries which could maintain service for a day or more if the commercial power went out. It was these emergency backups that enabled New Yorkers on that dark night in 1965 to call across the country while they were unable to take the elevator to the ground floor.

Today, thanks to the FCC's Toaster Theory mentality, the majority of PBX and key equipment installations are owned by customers. These have been installed by equipment manufacturers, often at bargain prices. But nothing comes free, and in many cases these installations have no emergency power—that would have cost more. So these customers have far less reliable service than they had before.

Surely, another winner of the Dubious Progress Award.

Of course, it didn't have to be this way. The FCC could have paid more attention to the panel of experts it hired after the Carterfone decision.

The commission had asked for comments from interested parties on the use of protective interfaces. After being inundated with responses ranging from total acceptance to complete rejection, the FCC turned to the National Academy of Sciences' Computer and Engineering board in September, 1969, and asked for a study on the technical factors involved with customer attachments to the telephone network. Lewis S. Billig of the MITRE Corporation headed the panel of fifteen technical leaders from companies in various parts of the electrical communications field, including two AT&T executives.

In June, 1970, the panel issued its report. It concluded that the network could be harmed by uncontrolled interconnection, specifying hazardous voltages, longitudinal imbalances, signals of excessive power or improper spectrum, and improper network control signaling as potential problem areas. However, the panel also stated that a properly designed program of equipment standardization along with enforced certification could safely permit customer-owned equipment to be connected to the network without a protective interface device.

(John deButts was to argue in 1974 that, "No system of certification we can envision, and no interface equipment, can provide a fully adequate alternative to the unequivocal and undivided responsibility for service that the common carrier principle imposes.")

Unfortunately, the panel's charter was too narrowly defined and time for study was limited. It was asked to study the "harm to the network" issue and protection against such harm, nothing more. The FCC didn't want to know about cost, reliability, quality of service, and economic policy. No. Harm was the issue, and that was that. It effectively tied the hands of the panel. Still, the panel did cite instances of increased trouble reports and network "traffic" jams caused by customer-owned equipment, though it didn't address them since it considered that issues involving traffic engineering were outside its assigned scope. Some members of the panel also expressed concern that interconnection might hurt innovation in the network. But the final report clearly stated that these other factors were not evaluated and urged the FCC to consider them carefully before making any final judgments.

For whatever reason, the FCC ignored the panel's caveats on cost, reliability, and service quality. Instead, it focussed only on its conclusions that certification of equipment was needed to prevent harm to the network.

But what of harm to the customer when quality of service was shunted aside?

The Public Utility Commission of Canada's Prince Edward Island addressed that very issue in a lengthy investigation of the effects of interconnection of user-owned equipment.

Its conclusion? Prince Edward's Island's Public Utility Commission rejected customer ownership. It said that there would be a massive deterioration in the quality of telephone service and that interconnection would be "immoral, economically unsound, and not in the public interest."

15

A Conspiracy
of Silence

A National Academy of Sciences panel yesterday warned that uncontrolled use of customer-owned telephone equipment could harm the nation's telecommunications system.

In a recently completed nine-month technical study, the group recommended that standards for such connections be set and strictly enforced to protect telephone service.

—Washington *Post*
July 6, 1970

WASHINGTON—Secretary of Defense Caspar Weinberger yesterday asked the Justice Department to drop its antitrust suit against AT&T.

In a letter to Attorney General William French Smith, Weinberger said the breakup of the Bell System would greatly jeopardize national security.

The Defense Department maintains dismantling the Bell

System, as called for in the suit, would seriously harm U.S. defense communications.

—Louisville *Courier-Journal*
February 22, 1981

COMMERCE SECRETARY SAYS BELL BREAKUP WILL
BRING BIG TRADE DEFICIT

WASHINGTON—Secretary of Commerce Malcolm Baldrige yesterday said the breakup of the Bell System would lead to an annual trade deficit in telecommunications equipment exceeding $1 billion.

In urging the dismissal of the federal suit against AT&T, he added that Japan would be the main beneficiary if AT&T were dismantled.

—San Francisco *Chronicle*
May 15, 1981

"All I know is what I read in the newspapers," humorist Will Rogers once said. He wouldn't have known any of the above information, because the stories never ran.

Have a television evangelist fall from grace in a motel room and it's front page, bannered across the top, hot stuff. Let a presidential candidate get caught with his pants down—figuratively, of course—on some Bahamian yacht, and the wire services are blasting details around the world. Watergate, Abscam, Irangate, presidential intestines, all big news and covered ad nauseum in newspapers, magazines, and on the nightly network news.

The nation's press devoted seemingly endless pages and hours of coverage to the possible mishandling of $15 million in funds intended for Nicaraguan "contras," and yet there was hardly a whisper or a trickle of ink expended on a loss of several thousand times that as a result of the calamitous butchering of the Bell System.

If Will Rogers were alive today he might do well to ask why he wasn't allowed to know about the economic and technical blunders that led to telephone network competition and to the breakup of the Bell System. Why was this whole horrendous process practically kept a secret for more than twenty years? Where was the press? Why wasn't it pouncing on this multi-billion dollar ripoff of the public? Why wasn't it analyzing, dissecting, explaining? Where were the review-and-opinion pieces presenting the various sides of the issues?

One can only wonder if there wasn't some kind of conspiracy, a conspiracy of silence. For that is what there was—a harmful, negligent silence.

Unfortunately the public never learns. Despite frequent, spectacular revelations of government corruption, venality, and wrongdoing, the American people cling strongly to their naive belief that their government leaders are inherently honest and intelligent.

The average John Q. doesn't want to entertain the thought that his "public servants" might have stolen billions of dollars from his own pocket so that a few wealthy entrepreneurs could line theirs. That a long succession of ill-conceived and stupid actions might have jeopardized our national defense, damaged our balance of trade, and crippled one of our nation's proudest resources—its telephone system.

All this would be inconceivable to John Q. unless—unless the media got on the story.

"Things don't get corrected until they're publicized" David Nyhan said in the Boston *Globe* in 1987. They don't even get noticed. The public depends on the press to alert it to crimes and malfeasance. The country is too large for information to be disseminated by word of mouth, over the back fence. Should the media ignore the Bell System's arguments about the high costs and penalties of contrived competition and divestiture, naturally the public will too.

Significant events such as Judge Richey's ruling in favor of AT&T were either buried behind the obituaries or ignored altogether. And there were no in-depth explanations of how MCI was able to undercut AT&T's prices on long distance service at the expense of the public and what the real cost to the public was.

What happened? Was the press caught napping? Was there some sort of incredible lapse in editorial judgment—the ongoing story was too dry, too complicated, too unimportant? It wasn't "sexy" enough? Or were there other factors involved?

One can only speculate.

For instance, it's a given that newspapers, magazines, television, and radio depend for their existence on advertisements—ads pay the way and make the profit. Keeping that in mind, it should also be remembered that competition in the telephone industry meant many new advertisers. All of a sudden there were MCI, Sprint, the Baby Bells, AT&T, equipment manufacturers—foreign and domestic—all competing for the market *and* competing for advertising space. It was surely the ad salesperson's dream come true.

It just can't be disputed there are far more pages of telecommunications advertising in the media now than before divestiture—the individual Bell companies, when part of AT&T, did little other than local advertising. Now all of them are in there touting their wares nationally.

While the media can insist there's a firm, impenetrable barrier between newsroom and ad department, and editorial content is kept separate, one can only wonder.

And then, if this silence didn't arise out of greed and coveting more advertising lines, might it have had something to do with the media's well-known liberal tendencies? Did the nation's journalists simplistically assume that big is bad, and all large monopolistic corporations natural enemies? That any informational release from a Big (shudder) Business should be tossed in the circular file cabinet as so much propaganda?

Or maybe the press didn't oppose the Bell breakup because of its own business interests. After all, most publishers foresee a bright future for Videotext type services. These have the capability of displaying and printing news and other information of public interest in the home and already have been extensively tested in various locations. Knight-Ridder publications and AT&T conducted one such trial in Coral Gables, Florida, in the early 1980s. Despite this collaboration, many publishers are afraid AT&T will get into this market as the producer of information as well as its transmitter.

It's been argued that AT&T's being allowed to extend its monopoly privileges into electronic publishing could have serious deleterious effects on First Amendment rights, that AT&T would then be in the position of limiting and controlling information. Thus, anything that would damage the Bell monopoly would work to the advantage of the publishers.

But, of course, this argument is no longer debatable since Judge Greene has barred AT&T from entering the publishing field, at least for several years. In view of that, one again might wonder if it was a coincidence that Judge Greene and divestiture got very little opposition from the press.

Of course, AT&T didn't do very much to get more favorable press. Direct appeals to the public through advertising were scarce. The company did get cranked up one time, when it was lobbying for Congressional passage of the Consumer Communications Reform Act in 1976. But AT&T's Chairman deButts approached the lobbying effort with a heavy hand and an insistence that almost all terminal equipment competition be eliminated. As a result, the effort backfired, and created more antagonism in Congress than support. The press apparently had little desire to be associated with what was clearly a lost cause. There was no outpouring of editorials supporting the Act.

Before the Bell System was broken up, *The New York Times* ran a poll that revealed 80 percent of consumers were satisfied with their phone service. No other business received such a high satisfaction rating. But despite liking the service received from the Bell System, there was no outcry against divestiture. If as little as one percent of the public had protested with letters to the editor or to their lawmakers, they could not have been ignored.

But there was no such protest. Obviously the public had no idea what divestiture was sure to bring—poorer and more costly service. What has happened could not have been perpetrated without a conspiracy of silence by the press that was supported actively or passively by the FCC, the Justice Department, state regulatory commissions, and indirectly even by AT&T.

Not that everyone was silent. There were individuals, author Constantine Kraus included, who tried over the years to alert the public to what was happening. Articles were submitted to leading publications, columnists, and government leaders. But nothing came of them. One exception was an article, "Social Consciousness in Communications Engineering," that did see the light of day in the *Institute of Electrical and Electronic Engineers Communications* magazine, in May, 1976. The article correctly predicted the outcome of the emerging telecommunications policies of the government and accurately forecast how the public would be affected. It explained why long distance competition amounts to stealing from the poor to benefit the rich, and why competition in public utility operations forces the consumer to pay for the sum of the costs of all competitors.

If our esteemed government representatives read the article, they disregarded it, just as they disregarded the economic truths revealed in it.

It's a funny, if unhappy, state of affairs that the press often seems proud of its ignorance of technical matters. It's easier to get on bandwagons and go on technological witchhunts instead of really scientifically evaluating and understanding the issues. As a result, our society has been severely damaged. Consider nuclear power.

The expansion of nuclear power plants in the United States has come to a virtual halt while other nations are forging ahead. This is despite solid evidence that using nuclear power for energy is safer than fossil fuel and that many of the other options will be increasingly costly or not even available within the foreseeable future. The press has harped on what it perceives as the negative aspects of nuclear energy, the result being that this viable power source is being abandoned.

Then there have been public protests, often supported by the press, against possible hazards of microwave radiation in long distance telecommunications systems. These protests seriously delayed and added to the cost of several vital communications projects despite evidence that such radiation, which is present naturally in the atmosphere, is totally harmless.

And there are numerous other examples, with the introduction of competition into the telecommunications network and the subsequent breakup of Bell not the least of them.

The public needs to be educated on technological and economic matters. It's of vital importance to the future of our nation. And that education must begin in the press and other media. An enlightened and aroused public can then galvanize its executive and legislative leaders into educating themselves—and then, perhaps, into undoing some of the grave harm they have done.

16

The Cost
of the Crimes

What have we paid already and what will we pay in the future
for the

- Uncontrolled connection of terminal equipment
- Destruction of end-to-end responsibility
- Disregarding of the National Academy of Sciences panel's warnings and recommendations
- Introduction of contrived competition
- Failure of Congress to do its job and legislate
- FCC's Computer II decision depriving customers of modern services and stalling technological progress
- Forcing of local operating companies out of the business of providing telephone sets
- Destruction of the integrated network
- Fracturing of the Bell Laboratories
- Actions of Judge Greene that led to the network's current illogical, uneconomic structure
- Forcing of independent companies to struggle and compete where they used to work in harmony?

Crime is running a stoplight. Crime is a mugger going
bump in the night. Crime is a bookkeeper embezzling com-

pany funds. Crime is a president covering up illegal campaign contributions. And crime is an evil act, an act contrary to the public good.

There have been crimes against the country so immense, so expensive they seem almost malicious in intent. But Woodward and Bernstein have not reunited to investigate them, so it will probably never be known to what extent these awful decisions were motivated by good intentions, ignorance, spite, or greed.

The most serious of the crimes was divestiture. And the pity is it didn't have to be. Even if it was decided that the local operating companies should be separated from AT&T, why did it have to be done is such a ridiculous and costly manner by separating exchange and long distance service? Why couldn't the companies have been separately owned while retaining the integrated network? It could have been done through allocation of costs and revenues.

Then there would have been a "divestiture" without the serious decline in quality of service, the rise in local rates, or the billions of dollars lost.

But that's not how it was done and now the staggering price must be paid for these economic and technological crimes.

Here then, are the numbers, the costs of the crimes, the damage report in dollars and cents.

First there are the one-time initial costs for the telecommunications companies:

Actual direct costs of prosecuting and defending the antitrust case plus estimated administrative costs for the subsequent separation of assets	$0.9 billion
Cost of providing equal access ($20 billion by the local Bells and $2.7 billion by AT&T, but it must be assumed about half was for plant modernization that would have been spent anyway)	$11.3 billion

Capital confiscation resulting from the FCC's insistence on customer-owned equipment and the scrapping of Bell System terminal equipment—assuming $40 worth of equipment per residence customer with 100 million customers and about half this equipment made obsolete, adding another $2 billion for PBX and other business equipment	$4.0 billion
Unnecessary duplication of intercity transmission facilities (such as new fiber optic waveguide systems built by US-Sprint and MCI)	$5.0 billion
Unnecessary toll switching systems to comply with the settlement rules banning joint ownership (using costs in Pennsylvania and assuming total nationwide costs at 20 times this)	$0.7 billion
Operator services being duplicated by AT&T and the Baby Bells (again using the Pennsylvania costs and multiplying by 20)	$0.5 billion
Cost of plant built to replace that dismantled so that inter- and intra-LATA separation could be achieved. (Pennsylvania costs times 20)	$1.6 billion
Exchange bypass. These are ongoing costs, but to date are estimated at—	$1.0 billion
Total:	$25.0 billion

Twenty-five billion, not million, dollars. The national budget—and debt—sailing around in the ionosphere, may have inured many to these hard-to-imagine sums. It almost becomes unreal, Monopoly money to be tossed around without thought of actual value. But think of this $25 billion in terms of $100 for each man, woman, and child in the country. And that's only the one-time charges. Added to that must be the ongoing costs associated with duplicated operating and maintenance forces, costs of additional communications consultants needed to untangle the jumble, and costs of excessive administrative overhead.

How will these hurt the customer's pocketbook?

To add up the ongoing costs, the following assumptions—based on discussions with experts in the field—may be made.

- There are 200 million telephones in the U.S. with 150 million of them in 100 million residences. The remainder are business phones, half key systems, half PBXs.
- The average residence telephone line rate has increased 50 percent from $180 annually to $270. This included the FCC-mandated access charges.
- Residence installations have an average life of six years Fifteen percent of these new installations are in new homes and cost $250. Eighty-five percent are in existing homes and cost $50.
- Of the 500,000 PBXs, 150,000 are replaced or upgraded annually. Assume an average consultant's fee of $4000 per installation. Also assume that wrong choices are made in 25 percent of the cases and a different PBX has to be installed.
- Average costs of PBXs are $1100 per station. Average costs of key systems are $900 per station.
- There are 30 million business telephone lines in service and the rates for these have increased about $150 per line annually. This includes the FCC-mandated access line charges.
- Private line rates have approximately doubled, from $12 billion to $24 billion annually.
- The average annual charge factor (covering cost of money, depreciation, maintenance, etc.) for terminal equipment is 40 percent.

Using these assumptions, here is what the consumer is paying more each year.

ITEM	ANNUAL COST
Residence customers:	
Purchase of telephone instruments	
(150 million telephones × $50 × .4)	$3.0 billion
Increased residence line rates	
(100 million residences × $90)	$9.0 billion
New installations in new homes	
(100 million × .15 × $250 × .167)	$0.625 billion
New installations in existing homes	
(100 million × .85 × $50 × .167)	$0.71 billion
Residence Total:	$13.335 billion
Business customers:	
Purchase costs of equipment	
PBX—(25 million × $1100 × .4)	$11.0 billion
Key—(25 million × $900 × .4)	$ 9.0 billion
Replacing defective or wrong choice PBX	
(11 billion × .25)	$ 2.75 billion
Less rental costs prior to purchase	
(50 million × $100)	− $ 5.0 billion
Consulting fees	
(150,000 × $4000)	$ 0.6 billion
Increased business line rates	
(30 million × $150)	$ 4.5 billion
Special services rate increases	$12.0 billion
Business Total:	$34.85 billion
GRAND TOTAL:	$48.2 billion

Allowing for an estimated yearly savings of $3 billion on long distance calls, the customers are still paying in excess of $45 billion more—and that is every year—for telephone service.

There is no hope in sight. These costs are only going to climb. Each year. Higher and Higher.

Over a ten-year period, simple arithmetic shows that we're talking about $450 billion. To this must be added the $25 billion in one-time costs and another estimated $300 billion that represents reductions in telephone rates that could and should be made, but were not. Nor were customers' monthly charges reduced to reflect the fact that their telephone sets are no longer furnished by the company. They were also not reduced to reflect the continuing advances in technology that allowed companies to cut their staffs and otherwise lower their costs. All in all, we can point to a staggering and deplorable total of as much as $800 billion in excessive and unnecessary expenditures by American telephone customers.

And that's not pocket change.

All those are costs that can be measured in dollars and cents, but a thief can take a victim's peace of mind along with his VCR. With this crime, there was stolen the opportunity to make maximum use of our technological capabilities.

Service quality has been allowed to deteriorate.
Continuity and reliability of service have been degraded.
It's more difficult to obtain, use, and maintain service.
Billing is more confusing.
Regulatory authorities have failed to approve advanced service offerings, causing a virtual stagnation in progress for residence and small business users.
Jobs were lost to overseas manufacturers.
The nation's trade balance was adversely affected.
The Bell Laboratories was lost as a national treasure.
National defense was weakened.

But gee, look at the bright side, at all the consumer gains.

A wider—but not necessarily better—range of choices.
Lower—by about 35 percent—long distance rates.

It would be comforting if that list were a bit longer, but it isn't. For a few of the largest business customers, these gains may make up for some of the losses. But for the rest of the population? As W. Brooke Tunstall, former AT&T vice-president put it in 1985:

"The ordinary American telephone user has been blindsided. . . . Divestiture's supposed purpose was to benefit consumers. But it is difficult to find a consumer who is grateful for the favor. ... A phenomenon noticeable throughout history is the pursuit by government of policy contrary to the welfare or advantage of the body being governed. Surely a prime example in our time has been the breakup of the Bell System."

Amen, W. Brooke Tunstall. Amen.

17

One Communications System for Mankind

There's a future out there, a future in telecommunications that will reshape and redefine our lives. Our children's children will question how we survived in our Dark Ages era, much as we wonder how civilizations got on without indoor plumbing, electricity, or knowing the earth was round.

There has been a tendency to underestimate future technological capabilities and progress. Ability outruns imagination. It's hard to believe now in the late 1980s what one of the authors, Alfred Duerig, heard at a Bell Laboratories lecture a mere thirty years ago.

The speaker was John Shive, who had participated in the invention of the transistor. He speculated that by the last decades of the century transistors might cost as little as a cent each. A cent! Even he seemed to only half believe the notion.

As it turned out, the cost dropped much faster and more dramatically than that because of large-scale integration of circuit components. Transistor costs in integrated circuits are now measured in fractions—even hundredths—of a cent.

More recently, technology outstripped projections in the growing field of fiber optics and related projects. The U.S. Department of Commerce, in 1981, predicted there would be a

$1 billion market in this field by 1990. That figure was sur-
passed only four years later and by 1986 *Time* magazine was
forecasting a $50 billion market size by 1990.

Time and again, progress in telecommunications technology has exceeded even the most optimistic forecasts. But
there is more to progress than technological advances. The
ability to manage and apply technology is essential. All too
frequently regulatory and judicial restraints or inept marketing decisions—as in the case of Centrex—have left technology standing idle.

The real challenge is in unleashing and managing technology that to a large extent is already available. Then and only
then can the nation enjoy the fruits of its fabulous future in
telecommunications.

Incredible progress has been made already. Those who
were around in 1945 remember all too well what it was like
trying to place a coast-to-coast telephone call. Delays of several hours were normal. Operator connections were made
manually and were subject to human disruptions. And by today's standards, transmission quality was very poor. Technological advances during the last forty years have made a call
across the country as clear as one down the block.

One advance was the high capacity transmission systems
that became available immediately following World War II.
Using coaxial cable and point-to-point microwave radio, a
vast nationwide analog transmission and switching network
was built. These developments opened the way for direct distance dialing.

Direct dialing did much more than speed up the time it
took to get a long distance call through. It also led to dramatically lower long distance charges because operators were no
longer needed. Once it became cheaper to call Cousin Molly
in Minneapolis, more people started calling. Long distance
was no longer a service for the rich. More calls brought in
more revenue. The Bell System then had more money to subsidize local operations and could therefore finally reach its
goal of universal service.

The universality of the telephone and the ease of dialing
calls had profound sociological effects on the nation. The tel-

ephone replaced the letter as the primary means of communications. Many people virtually abandoned mail for most business and social correspondence, using the written word only when a record was required. This phenomenon is true only in the United States and Canada.

The more recent invention of optical fiber waveguide is a breakthrough of the same magnitude as microwave radio. Fiber optic transmission will have incredible impact on both the home and office by making available unlimited amounts of information.

With sand as the basic ingredient of glass fiber, it should become—literally—dirt cheap. Progress in this direction has already been impressive.

In 1982, the average cost per meter of optical grade fiber was $2.82. By 1987 it had dropped to $0.40, and it will surely fall even further.

Along with these spectacular advancements in high capacity transmission systems came corresponding improvements in switching technology. These began in 1948 with #5 Crossbar. This was a local end office switching system using early computer technology to provide centralized memory and logic functions electromechanically. Using similar technology, the #4A Crossbar system provided interoffice connections, making nationwide long distance dialing possible.

The #5 Crossbar was succeeded in the late sixties and seventies by analog electronic switching systems, with the #1A ESS for local exchange service being the most advanced example. In this, transistors and similar devices and computer software took the place of electromechanical switches and relays. In the 1980s, digital switching systems such as Western Electric's #5E and Northern Telecom's DMS-100 in turn have replaced the analog systems in new installations. Meanwhile, the enormous capacity of Western Electric's #4E digital electronic long distance switcher allowed a single one of these machines to replace two or more of the older #4A electromechanical installations.

Great strides were also made in signaling and supervision. These are the non-voice transmissions necessary to establish and monitor connections such as dial pulses, on- and off-hook signals, and the like. Several generations of these supervisory systems used in-band signaling, which means these functions were an integral part of each circuit. Now such out-of-band systems as CCIS and CCITT #7 provide totally separate channels for supervisory and dialing functions. This allows for faster connections, lower costs, and protection against fraud by "phone phreaks" using the so-called "blue boxes." They also allow access to large central databases for number translations—for example, those toll-free 800 numbers, which get translated to standard ten-digit numbers that can be switched through the network—as well as a variety of other services, many of which are still not available to the public.

Long distance telephony will only improve in years to come. A recent decision to route calls using a system called DNHR (Dynamic Non-Hierarchical Routing) plus a reduction in the number of switches in a connection will improve network efficiency by at least 10 percent. Then connect time—the gap between the end of dialing and the start of ringing—should ultimately be about a second, down from the twelve seconds of 1970.

Several technologies will be indispensable to the ever-expanding future of telecommunications. Certainly one of them is solid state electronics, which includes the transistor, LASER, integrated circuitry, microchips, and a host of other devices. Another is digital switching and transmission technology, by which information of all types is reduced to a string of binary ones and zeros and handled entirely in that form. A third is the concept of stored program control, embodied in computers and in all modern telephone switching systems.

While it might seem that only engineers and scientists would get their pulse rate racing over all this, it opens the door to the future for the entire country—and world. But to fully realize the potential of what's available now and what will be available, goals must be set. And a primary one should be making the telecommunications system available to everyone, in other words, a "democratization of communications."

To put the future in perspective, a brief look backwards is in order.

Man always needed to communicate with his fellow man.

It was fairly simple in the beginning. One merely turned to the next person and grunted. But when face-to-face conversation was impossible, other means had to be developed. The cavemen started it all with their wall drawings. The Egyptians had their papyrus. But that still left the problem of getting the message from one person to the next when they weren't sitting side by side.

The Egyptians solved that with a mail courier system as early as 2000 B.C. The Chinese, Greeks, and Romans set up their own later on. But these early postal systems were for the elite only. The ordinary man had no access to them.

It wasn't until the mid-eighteenth century that Benjamin Franklin, as deputy postmaster general for the colonies, decided that a major priority was a universal communication system. The result was a mail service using horse-drawn vehicles that delivered written material to every business and home from Maine to Georgia.

In Western Europe, various systems of mail delivery were used. In the early 1840s a revolution took place, a revolution which made mail a government monopoly and opened up service to everyone at a moderate price. It was the British "penny post." Before the penny post, it was the recipient who paid for the delivery, not the sender. He was charged the sum of the charges of all the couriers involved. The cost of transporting a letter in Great Britain had averaged a shilling and a half, or about 35 cents.

Then Sir Rowland Hill came up with the idea that it should be the sender who pays for the delivery by buying a stamp to be affixed to the item being sent. The implementation of his concept took ten years and changed mail service from a disorganized profitless operation available to a few into an efficient, fast, low cost, profitmaking service available to everyone. His plan was soon adopted worldwide.

It democratized the communications system of the day. This basic principle of providing mail service for all must be applied today to electrical communications. Just as Franklin determined that every citizen should have access to the postal system, now everyone should be able to use its electronic equivalent. To accomplish this, political, technical, and economic barriers have to be eliminated.

The enormous strides in transportation have transformed

the world from a collection of farflung places to one big neighborhood. A worldwide universal system of communications is essential for this new, smaller globe. To achieve this, the best features of the systems already in place—mail, telephone, television, computers—must be integrated. It is necessary to design and create a more comprehensive, efficient communication system that will interact universally between men and machines.

For some time business users with PBXs and other large systems have had modern services to speed their communications and make them more efficient. These have included conference calling, priority calling, centralized answering services, and more. Small businesses and residence users have been denied similar advances and not because of technological limitations. The technology is there. The bottleneck has been bureaucratic and regulatory. As Colonel Bolling put it in his paper, "AT&T—Aftermath of Antitrust," "In telecommunications, technology has leaped while federal law has crawled."

These bureaucratic and regulatory obstructions have seriously impeded economic and technological progress in information transmission.

Where to start?

With the basic telephone system itself.

Despite remarkable technical advances, it is still a woefully inefficient system. Bearing in mind that its basic objective is to transmit a message from calling party to called party with a minimum of effort as quickly and accurately as possible, consider the following:

1) Out of some 600 billion attempted calls made annually in the United States, only 420 billion—or 70 percent—get through on the first try. The other 180 billion are equally divided between "busies" and "no answers." With the present system, most of the 180 billion unsuccessful callers have no alternative but to try again.

2) The chances of getting through on the first try to the specific person being called is about one in three. A call to a professional person, such as a doctor, has only a one in twenty chance for completion on the first try.

3) Even when calls are completed to the desired party, much time is wasted because, in about half the cases, two-way conversation is unnecessary. All the caller wants to do is deliver a message and not necessarily engage in conversation.

One way to make the present communications system more efficient would be to stop calls from being "lost" unless the receiving party chooses to reject them. There are several steps which can—and should—be taken to achieve that goal, and little new technology would be needed.

The first would be to provide universal touchtone dialing. This is an essential improvement, not only to save dialing time, but because many of the new services will not work with dial pulses.

It would be an easy change since all electronic switching systems have tone capability, and by 1990 more than 98 percent of U.S. telephones will be served by electronic switching. In electronic offices, touch tone calls cost the telephone company less to handle than dial pulsed ones.

One stumbling block would be the customer. He has to have a touch tone instrument. As it is, he may elect to stay with his old pulse dialing phone because he doesn't want to pay extra for touchtone telephone service.

This premium charge makes no sense since the telephone company as much as the customer benefits from getting a call completed faster. If a unified Bell System still existed and still owned all the terminal equipment, there would be a gradual programmed phaseout of dial phones. In fact, just such a plan was proposed by Bell Laboratories more than a decade ago, with a projected completion date of 1990.

As it is now, the customer manages his own equipment modernization. To nudge him into replacing his telephone will take an incentive such as lowering his line rate or selling him inexpensive touchtone instruments.

To date only Utah's state regulatory commission has required that the premium charge be dropped from touch tone service. Other commissions, such as New York's, have denied the telephone company's request to eliminate the additional charge, saying these extra revenues should be used to subsidize basic services.

Carl Oppedahl, a New York lawyer specializing in technological litigation, disagreed in a 1986 *Wall Street Journal* Op-Ed piece. "... Forcing local phone companies to charge extra for touch tone promotes inefficiencies [such as] needlessly expensive central offices. ... State utility regulators should do the efficient thing and drop their present requirements that tone be charged for."

Another step that should be taken for a more efficient system would be to provide call waiting on a universal basis. Again, no new technology is needed here. Electronic central offices are already offering this service for an additional monthly charge.

But there is no need for that charge. The cost of providing the service is quite small and is actually outweighed by the potential benefits to the telephone company. Call waiting would eliminate a large portion of the annual 90 billion busy signals. When a customer gets a busy signal, it ties up the network without providing revenue to the telephone company. The telephone industry should consider providing this service without extra charge.

One independent telephone company has already included touchtone service, call waiting, three-way calling, speed calling, and call forwarding as an integral part of its basic service.

Centralized answering and recording facilities should also be available to everyone. Going out? Instead of turning on an answering machine, the customer would dial a code to activate the service. All incoming calls would then be routed to centralized answering equipment. The caller (or perhaps certain selected callers) would hear a pre-recorded message and an invitation to leave his message.

The customer could retrieve his messages at any time from any telephone by dialing his private code. The telephone company would get the revenue from a completed call, whereas now it gets nothing from the unanswered call attempt. That benefit to the company could be reflected in the customer's charges. A per-time charge is appropriate but could be set below the telephone company's actual costs providing the service.

So why not continue using individual answering machines as many people do now? For one thing, they don't get used that often. Highly efficient centralized equipment makes far more economic sense. Add to that the reliability and higher quality service provided by centralized equipment and it's clear that denying customers this option is analogous to outlawing central heating in homes and requiring individual furnaces in each room.

There is still another service needed to improve the network's efficiency—caller identification.

Imagine. The phone rings. Before picking up the handset, the customer glances at a display unit (provided for a charge by the phone company). On it he sees the number—maybe even the listed name—of the caller. He can then decide to answer or not. If he does, right away the call can be shortened since the caller doesn't need to identify him or herself. Also, junk calls, that persistent insurance company that loves to tie up the phone with a computer-generated "Hi! We understand you've just bought a new house!" can be ignored.

Versions of this service have already been test-marketed for several years in Harrisburg, Pennsylvania, and Orlando, Florida, and have been well-received by the public.

An alternative to the display unit, which would have to be installed, would allow the use of the standard telephone the customer already has. This would have a computer-generated message announce the calling party's identity when the customer lifts the handset. To give the customer a chance to decide whether to take the call, the central office would have to continue to return an audible ringing signal to the caller. If the called party then wishes to speak, he could stay on the line or flash the switchhook. If he doesn't wish to accept the call, he could key in a special code to return a "no-answer tone."

This "no-answer tone"—please let someone come up with a better name for it soon—is still another new feature that should be made available to all subscribers. With this service the customer will activate a tone by dialing a code from his telephone when he's leaving the house or just doesn't want to answer his phone.

Then instead of getting a ringing sound, as he would now, a caller will receive the no-answer tone and then can hang up immediately. This will save him precious seconds of his own time as well as the network's time and cost. To deactivate the no-answer tone, the user would only have to lift his handset.

Of course, there is always the possibility of the customer coming home and forgetting to deactivate the "no-answer tone." That could be circumvented by a short signal being sent to his telephone on every incoming call attempt. That would give him an audible indication his telephone was still rejecting calls.

While this service would be easy and inexpensive to provide, it basically benefits the telephone company and the caller, but must be activated by the call-receiving party. To make it desirable for the called party to use, financial incentives could be provided. He might receive a credit on his bill based on the total time per month the no-answer tone is activated. Or, a credit could be based on the relative number of times a no-answer tone was sent from his line as compared with ringing and no answer. Either of these methods should be feasible with modern recording and billing equipment.

All these services could be furnished inexpensively and fairly quickly on today's network if desired. But the question becomes, is it desirable? Are there more important issues than mere customer comfort and "Gosh, wouldn't that be convenient to have?"

A look at some dollar figures might answer that. It's difficult to appreciate the value of a few seconds saved here and there until those seconds start adding up—into the billions.

First some assumptions that can be made.

> 600 million telephone call attempts are made annually in the United States
>
> 90 billion of these get busy signals
>
> 90 billion get no answers
>
> 13 seconds are required on average to dial a call on a rotary dial
>
> 6 seconds are required on average using a touchtone
>
> 10 seconds elapse on average from the end of dialing to beginning of conversation
>
> 20 seconds ringing time elapses on no answer calls before hangup

Using these assumptions, the five services can be analyzed in term of time saved.

1) Universal touchtone—Assume that approximately one-fifth of calls made today do not use touchtone, then converting these to touchtone would save:

600 billion × .2 × (13 − 6) = 850 billion seconds

2) Call Waiting—Assume that this would eliminate callbacks on half the 90 billion busy signals now encountered (in the other half the called party will choose not to answer) and that dialing savings on callbacks are 8 seconds based on 75 percent of phones being touchtone. Connection time is 10 seconds. The savings become:

90 billion × .5 × (8 + 10) = 810 billion seconds

3) Centralized Answering—Assume this is priced low enough so 50 percent of customers will use it. Take the 8 second average dialing time and 10 seconds connect time plus 20 seconds ringing time. Add to that 45 seconds saved per call because of briefer, more concise messages. The savings become:

90 billion × .5 × (8 + 10 + 20 + 45) = 3735 billion seconds

4) Caller Identification—Assume 50 percent of called parties use this, and they choose not to answer the call 10 percent of the time. Also assume that 10 seconds are saved by not having to identify the caller and 30 seconds are saved in identification and conversation on rejected calls. The savings:

(420 billion × .5 × .90 × 10) + (420 billion × .5 × .10 × 30)
= 2520 billion seconds:

5) No answer tone—Assume 50 percent of all nonanswered calls go to telephones with an activated no-answer tone. With the ringing time on an average nonanswered call being 20 seconds, the savings become:

90 billion × .5 × 20 = 900 billion seconds.

The total savings on all five services is 8815 billion seconds each year. This amounts to 280,000 years. In addition to that the reduction in network holding time must also be considered. Putting a dollar figure on the savings has to be inexact, but say

—The customer's time is worth $10,000 a year on average;

—And the usage-sensitive portion of the telephone network investment is $20 billion and the average holding time of the 420 billion annual completed calls is 120 seconds. Using an annual charge on capital of 20 percent, the network cost of one second of call holding time is $.00008.

Using these figures, customer savings can be calculated at some $3 billion annually and the network savings at $.7 billion.

Even if these figures are too high by a factor of ten, the savings would still fully justify the modest investment needed to bring these services into operation. And this is without considering the beneficial impact that speeding up telecommunications would have on the business world.

Another way to look at the benefits is to consider the percentage of total time that the customer spends at the telephone. This amounts to a 25 percent reduction in time while improving the completion of information transmission. The possibility of another 25 percent reduction may be secured by the auxiliary services described later in the chapter.

These five services are, at the very least, what should be offered—and soon. Beyond them, there are many other services that could further enhance and generate more savings in the communications network. Of course, in considering them, existing regulatory, legal, and organizational barriers have to be momentarily disregarded.

VOICEMAIL

In 1979, Bell Telephone Laboratories developed the 1A digital Voice Storage System, designed to function with electronic central offices and provide a group of features called "Custom Calling II." These included centralized answering and recording, not unlike what has already been described, and something known as "advance calling," or what Constantine Kraus dubbed Voicegram service when he patented it in 1973. This is now universally known as Voicemail.

With Voicemail, a customer can have a prerecorded message delivered at a specified time.

For example, say Bill and Mary are going on an extended trip at the same time several of their friends will be celebrating birthdays. Bill and Mary can dial a code instructing the central office to forward pre-programmed voice greetings on the proper days.

Or Bill, in Los Angeles, wants Joe, in New York to get a message by the opening of the next business day. But Bill doesn't feel like getting up at 6 A.M. local time to have the message arrive at 9 A.M. east coast time. The solution is a Voicemail call timed for the 9 A.M. delivery.

What if Jim wants to call a meeting of his six department heads for next Thursday? He knows they'll be hard to reach and Jim doesn't want to waste time making a lot of calls to track them down. A single multi-addressed Voicemail call will do the job. Repeated attempts would be made automatically to deliver messages if the recipient fails to answer his phone on the first try. If he wishes, Jim can dial a special code for a status check. He could then find out which of his messages have been delivered.

At considerable cost, equipment to handle these services was installed at central offices in Philadelphia and Dallas. While awaiting FCC approval, telephone company employees enthusiastically partook of the services. But the FCC approval never came.

Then as part of its Computer II inquiry, FCC decided that these were enhanced service offerings. Under the new rules of Computer II, enhanced services could only be provided by a fully separated subsidiary. Physically, there was no way to separate the equipment from that which furnished basic telephone service since they worked over the same lines. Which is why no one has received an invitation from the telephone company to sign up for this new superduper service. Several millions of dollars worth of installed equipment was removed and junked.

In March, 1988, Judge Greene eased his restrictions considerably. He ruled that the Baby Bells could now offer such services as electronic mail, voice message services, and audiotex. The companies still cannot originate data, but as Judge Greene put it, granting "wide flexibility to the regional companies with respect to transmission systems and voice storage applications [will] bring this nation closer to the enjoyment of the full benefits of the information age." However, FCC approval is still required before these services can be offered.

There is no doubt Voicemail would go over well with the public. PBX users who already have something similar have shown it to be very popular. Voicemail one-way message service seems to have every advantage over the common telegram—speed, verbatim and personalized messages, controlled delivery time, and lower cost. One study projected that there would be 15 billion Voicemail messages delivered annually. The combination of Voicemail and centralized answering and recording should dramatically reduce the number of two-way conversations.

MAILGRAM

Western Union, in cooperation with the Post Office, is already supplying this service. A person wishing to send a short written message telephones Western Union and dictates his message to an operator who displays it on a screen. The operator then transmits the message to a printer at a post office near the recipient, who gets it delivered in the next-day's mail. This service should be expanded to permit direct customer connection to the network when he has the required data transmission and terminal capabilities.

A future enhancement to Mailgram might provide the capability to translate from one language to another. In addition, at the terminating end, the electrical data can be translated into speech and delivered anywhere in the world as Voicemail.

FAXGRAM

It would start in the post office. A person would go there with his graphic material that he needed to get to his business partner two cities down the pike. He pops some coins—or inserts his credit card—into a Faxgram transmitter. Zap. A copy of the material pops up at the post office two cities down the pike, and his business partner gets the material delivered in the next day's mail.

If the next day's mail isn't soon enough, the person would have the option of paying more for the material to be specially delivered.

Large corporations could bypass the trip to the post office

by having their own facsimile machines with direct access to the network.

It's estimated that Faxgram would cost 50 cents to a dollar per page, with a discount if more than one page was going to the same destination.

A similar service has been available in several European countries for years. It's not gotten started in the United States because of the arbitrary requirement that voice and record traffic must be separated. If Faxgram ever does go into operation, watch out Federal Express and those other courier services offering next-day door-to-door delivery at $20 an item.

The logical step after Faxgram and Mailgram would be televideogram, which would deliver the message or graphic material to the recipient's television—storing it if the TV is off.

INTEGRATION OF TELEPHONE AND TELEVISION

This idea of integrating the telephone and television isn't some 25th century Buck Rogers/Star Trek dreaming. It's already happening in Great Britain, under the name of PRESTEL, and in other European countries. It has also been tested in the United States.

The services that could be offered are limited only by the imagination—airline and train schedules, movie and other entertainment information (reviews as well as schedules), telephone voting, computer games, shopping services, libraries, investment data, and access to any sort of information that can be stored in databanks. Many home computer owners are already exploring some of these services using various commercial databanks.

Why there's no PRESTEL in the U.S. is largely due to political roadblocks. There is a question of how to keep separate the providers of communication services—regulated telephone companies—and providers of information—unregulated newspapers, broadcasters, etc. One solution might be to offer these as regulated public utility services that with some intelligent planning would assure any information provider full access to the network.

This kind of regulatory barrier is what led John R. Pierce, a former Bell Laboratories scientist and author, to say, "I fear that the future of communications in this country may be shaped more by regulation and legislation aimed at general social goals than by the potentialities of communications science and technology."

DIRECTORY INFORMATION

The phone directory, white and yellow pages, is nothing more than a huge databank. Today when someone can't or doesn't want to get to a directory, he calls directory assistance. The operator keys in the first few letters of the desired name and the various possibilities pop up on a display monitor. The source of the information viewed by the operator is an enormous computer memory, or database.

It would be fairly simple to bypass the operators and give customers—with the proper kind of alphanumeric keyboard and display terminal—direct access to this database. This would open up all sorts of possibilities. Instead of being restricted only to telephone number information, the directory service could be expanded to furnish zip codes and addresses of businesses even when the city and state are unknown, and so on.

A customer would be charged for holding time, which would vary depending on how complicated the request was. France already has such a service. It has a national databank with information on 23 million subscribers (a total of 25 billion alphanumerical characters.) Customers can gain access to this databank by entering complete or partial names—even ones that are misspelled—address information only, or the profession or service desired. The search can be within a small geographical area or expanded to cover the whole country.

To date, this is the world's most successful videotext system. To get people interested in using the various services, more than two million terminals, called Minitels, were initially made available to customers free of charge.

This French system handles some 35 million calls monthly, using the largest packet switching network in the world. And this in a country that 20 years ago had a telephone system that was a global joke.

BUYER CREDIT SERVICES

The Sharper Image catalog arrives. There on page 15 is *it*. The Abominator™—"only this board lets you do all five known sit-up exercises." And for only $129.95. The person must have it, right away, immediately. He grabs his credit card and calls the handy tollfree 800 number.

And then everything slows down considerably. He has to recite his name, address, credit card number, date of expiration, and what he wants. Then he has to listen as the operator reads all the information back which usually he then has to correct. And by the time it's all over, he's looking down at the catalog wondering if the Abominator™ is really going to give him a lean flat stomach, and whether the whole thing was more trouble than it was worth.

The ordering-on-credit process could be considerably shortened with a new service. This would transmit the caller's name, address, telephone number, and full credit information automatically to a display terminal in front of the order taker even before the telephone is answered. All that would be left to do would be tell the order taker what was wanted. Obviously, dispensing such information must be under the control of the caller, who would activate the service by dialing a special code along with the number being called.

These are but the barest tip of the communications service iceberg. So much more could be made available.

Electric and gas meters could be read over telephone lines.

Electrical appliances—furnaces, microwave ovens—could be turned on and operated remotely.

Banking and paying bills could be taken care of by phone.

Racetrack and lottery number bets could be placed.

And so much more. The only limit is the imagination.

As it is, restricting the vast computer network of the telephone to mere conversation is a gross underutilization of its potential.

By opening up the network to the future, our basic lifestyle can be changed—for the better. With computers and communications virtually eliminating the need for paperwork, many employees wouldn't have to commute to a high-rent downtown office or out to a distant industrial park. They'd be able to throw a robe on each morning, grab a cup of coffee, do their work and earn their pay at a video terminal in their own home. The only thing missing would be cookie breaks around the old water cooler.

Working at home means saving commuting time, working flexible hours, making home management and child care fit much better into the demands of the working world. Several companies have already begun experimenting with at-home employees and report a high degree of acceptance.

There would be great changes in education, as well. It could be virtually democratized. In 1988 it costs as much as $20,000 a year in tuition and board at an Ivy League school. $20,000! Even with taking second and third mortgages and hocking the family heirlooms, few families can afford that kind of expense. And even the less expensive schools are out of reach for most people.

Costs can be greatly reduced and availability of the best teachers increased by proper use of communications technology. It's already happening at Lehigh University where a fiber optic network was established in 1987 to link its classrooms with three AT&T-Bell Laboratories locations. This enables Laboratories' employees to take graduate-level courses at their worksites. It is the first fully interactive two-way audio-video educational network in the United States.

To bring this future into the present there must be a unified communications system—postal and electronic information must be completely complementary services. Even then an integrated system must be of superb quality to perform all these tasks, plus have the flexibility to take on new functions as required.

It must have the capability to accept information in printed, handwritten, or spoken form and to translate it first into electrical language and then back into something the recipient will understand. It should accept data at any speed, with protocols—a procedure or set of directions to provide the orderly flow of information between system devices— converted as necessary, and then transmit the data rapidly. The types of information that would be handled are:

1) voice messages (one- or two-way)
2) printed messages
3) data at various speeds (one way or two way)
4) one-way broadcast programs (audio or video) such as cable news networks.

To be able to perform these various functions, a universal communications system will need some basic characteristics. These include:

1) electrical connections completed in less than one second except for entertainment programs
2) transmission capabilities up to six megabits or more per second
3) store and forward capabilities for all voice and data services
4) facsimile equipment for conversion to data form and transmission of pictures, graphs, maps, etc.
5) access to libraries and data banks
6) available mode translation—print to data, data to voice or print, and eventually voice to data or print (optical readers that convert print to data and devices that translate English in data form to speech have already been developed)
7) within a few years, the capability of translating between languages should be added. (This would be accomplished by converting speech into digital data form, as is being done in Canada where weather forecasts are automatically translated between French and English)
8) access to directory data and conversion between forms—name and address to telephone number and vice versa, street address to telephone number and name, etc.

These requirements are only a beginning. Detailed studies of the public's communications needs must be conducted. Then an organization must be established and plans developed for meeting those needs.

Much of what has been described here depends on fast, accurate, and reliable transmission of information in voice or data form between two points. The existing intercity analog network is capable of meeting many of these requirements. However, the problem in making more use of this network has been in the high cost of terminal equipment and modems and with problems in local loops (the connections to the customer's premises). Furthermore, terminal equipment has been designed for the largely analog voice network and that places limits on what can be done. The answer isn't to redesign the terminal equipment. The answer is a whole new network—the ISDN, Integrated Services Digital Network.

There exists the capability now to convert all forms of information—again, voice, data, graphics, pictures—into digital form and to transport it in this form. It would seem logical, therefore, in this increasingly information-sensitive society, to consider building a single network to handle all these forms of information. Such a network is the ISDN.

Although many aspects of ISDN are being tested and there are some small networks in use, it will be years before a full-scale commercial network is in operation. It will be even more years before it will replace the analog and digital networks now in use.

Many firms are currently working on ISDN, and it is encouraging that an international effort is being made toward planning uniform standards and protocols. Full deployment of an ISDN will eventually link the whole world.

The ISDN of the future must have three key ingredients—uniform standards worldwide, user-friendliness, and a large bandwidth. End-to-end digital technology—all transmission and switching on a digital basis—can meet these requirements. But to make it work, the evolution of ISDN must be gradual and well-planned and in time should incorporate and replace existing switched and private networks.

At this point, the objectives of ISDN are much too limited. It should be appreciated that today only a small portion of the outside plant facilities are restrictive in frequency bandwidth transmission. Fiber optics with almost unlimited bandwidth are replacing intercity facilities and even the feeder cables in the subscriber plant. Only the distribution cable remains to be replaced by fiber optic technology.

It is interesting to note that ISDN represents the closing of a circle. The first electrical communication system, the telegraph, was, after all, a digital system. All information was reduced to on-off pulses of direct current. This gradually evolved into a network which was entirely analog. Now the road has curved back to a digital network.

What ISDN breaks down to is a bit stream of ones and zeros that stay just that regardless of the kind of information the ones and zeros represent. Ultimately there may be a standard wall socket in each room—as we now have electrical outlets—that would allow the user to plug in a telephone, computer, television set, facsimile machine, or any number of other devices, and instantly communicate around the world by voice, video, high speed data, or written word. The world would be reduced from a global village to a global living room.

It should be appreciated that ISDN is not a service but a technology. This technology has economic advantages which are large. However, the use of this technology should not impede the planning and implementation of the vast number of new services urgently needed today to greatly speed up the nation's communications.

There should be one communications system for mankind. But it won't happen without thorough planning and organization. Since the entire communications system operates as a single immense computer, it will have to be accepted that uniform standards and centralized control are needed. And that this computer should be controlled by a single entity or group of cooperating entities. Competition will not work.

And lastly, if this future is to be reached, the regulatory and legislative powers that be and will be must help and not impede progress.

Or we will be left behind, in the dark ages of communication.

18

Organizing for the Fabulous Future

"Who's in Charge?"
—Title of paper on emergency communications policy
by U.S. Air Force Colonel ROBERT REINMAN, 1984.

Nobody is in charge anymore, not really.

In one corner there's the FCC, which somewhere along the way lost sight of its function—to maximize quality of interstate telecommunications while minimizing cost. But even though it's been failing in its primary responsibility, it nonetheless has been moving into the territory of the state commissions in such matters as terminal equipment. The state public utilities commissions theoretically have the sole authority to regulate intrastate telephone business, and terminal equipment had always been subject to local regulation.

In another corner Judge Greene reigns, as he continues his supervisory role over AT&T and its offspring.

In corner number three, there is the Justice Department, which doesn't want to let go of its involvement in telecommunications issues.

Then there's a seemingly hapless Congress, forever trying

to assert itself into telecommunications and never quite succeeding.

And finally, Bellcore, free-floating with no real place of its own, assigned responsibilities but with no real authority to enforce anything.

It's like one of those "Find Five Things Wrong with This Picture," except in this case it's "Find Anything Right with This Picture." This patchwork has led to a lack of standards and deterioration of service. Modernization and reduction of costs are needed. Neither this nor any of the exciting possibilities for the future will be realized if sanity isn't restored to the nation's telecommunications.

Of course, that is easier said than done. A situation exists now where all the many actors are trying to beef up their roles—or get themselves written into the script in the first place. And naturally no one is interested in ending up on the cutting room floor.

The first step in getting some kind of order out of this chaos is education. Both the government and public need to recognize the blunders of the past. They must be told how lawyers and bureaucrats were allowed to take over a technologically complex telephone network that had been meticulously designed, built, and operated successfully and in the public interest for many years by scientists, economists, engineers, and managers. This scenario would work in an absurdist script by Ionesco or one of George Orwell's brutal satires, but not in real life.

All the perpetrators of these blunders should be brought before the public scrutiny, and that includes the press, which generally presented only one side of the story, thus preventing the public from knowing what was being done against its best interests. The lessons of history must be understood so we are not doomed to repeating the same mistakes.

Rep. John Dingell (D-Mich.), chairman of the House Committee on Commerce, in 1987 blamed a lot of the industry's problems on Judge Greene.

"I continue to be offended," he said, "by the fundamentally antidemocratic process whereby a single unelected unaccountable federal judge has transformed himself into a regulator without portfolio, arrogating the power to determine whether and when the American public will be allowed to receive the advanced services that are already available in countries with more enlightened telecommunications policies."

But in truth, Judge Greene is only one part of the problem. Right off the bat, there are two other obstacles to making any sort of progress in organizing for the future. One is the lack of centralized control of the industry, and the second is the insistence on competition.

Gene Kimmelman, director of the Consumer Federation of America, had this to say about competition in telecommunications. "The emperor known as Competition has no clothes. This is not the kind of competition that benefits the public."

Introducing competition has brought anarchy to the network. Where once AT&T set and enforced industry compatibility standards, now exchange plant and responsibility for exchange services is spread among the seven Baby Bells and twelve hundred independent companies. Each makes its own decisions based on a variety of interests which often have little to do with providing superb telephone service.

This fragmentation couldn't come at a worse time, when the nation is faced with a job of overwhelming size, complexity, and challenge. A single entity or cooperating—not competing—entities are needed to run the one vast computer which is our telecommunications network.

Many national leaders have regarded competition as some sort of economic panacea with the assumption that a free and unregulated marketplace will insure survival of the best. This is a nineteenth century concept not at all suited to today's realities that has been rejected by most modern economists.

The American Economic Association, in its founding statement, proclaimed:

> "We hold that the doctrine of laissez-faire is unsafe in politics and unsound in morals; and that it suggests an inadequate explanation of the relations between the state and its citizens."

For some reason, the obsolete hands-off principle of laissez-faire is seeing renewed acceptance and popularity with the argument that it is the only attitude consistent with individual liberty.

Nonsense.

How much individual freedom did post-war Germans and Japanese forego in their economic resurgences? Yet unrestricted competition was not allowed in those countries when it meant wasteful duplication and unplanned growth.

Some common sense has to be used if the nation is to build the ideal communications system for the future. And common sense dictates a return of central direction—such as AT&T used to supply—while acknowledging laissez-faire competition in communications for what it is—counterproductive and disorderly.

Of course, disorder and chaos are to be expected when no one and everyone is in charge.

Just take the issue of the Baby Bells entering other lines of business or expanding their telephone business by acquiring independent companies. The settlement agreement allowed this. But, also as part of the agreement, no such move could be made until the Justice Department had the opportunity to comment. Even then, Judge Greene had to give his approval—although late in 1987 he eliminated the need for his okay on nontelephone activities.

But wait, there are the FCC and state regulatory commissions. They're charged with assuring that the monopoly powers of the Baby Bells are not used to the disadvantage of the public.

All this overlapping responsibility does not work in the best interests of the companies or the public. All it does is create more work for lawyers, lobbyists, and bureaucrats. It costs the Baby Bells more money to deal with several different authorities, adding to overhead. Inevitably a higher overhead gets reflected in telephone rates. Uncertainty and confusion also make it difficult for the companies to establish long-range business plans and to concentrate on developing and applying technology needed for the future.

Ad hoc regulation by lawyer-bureaucrats doesn't work. It's failed today and will fail tomorrow. At present, no one agency—executive, legislative, or judicial—represents the public in communications policy. The public needs a champion. The question is how will it get one to lead it out of the existing maze.

But rather than wait for a hero to step forward and save the day, some other steps should be taken to clear up the mess.

What happened to the telecommunications industry was a scam, plain and simple. It was a scam perpetrated gradually and incrementally over some fifteen years. No single event—except perhaps the antitrust settlement itself—grabbed the headlines or became a "we interrupt the normally scheduled programming to bring you this" news bulletin. But it was far more important than many of the events that did, and will end up costing hundreds of billions of dollars.

All this makes educating the public and government leaders—one of the steps needed to right what has been wronged—even more difficult. But it must be done.

There has to be a clear understanding that competition in the telecommunications network is not only wrong, it's dishonest, just as any double-billing is dishonest. Competition has brought wasteful duplication and enormous costs, perhaps as much as $800 billion to date. And rest assured, there's more to follow. West Germany acknowledged in 1987 that competition in the telephone network is as wrong as it would be in electric, gas, and water distribution services by voting to retain its telecommunications network monopoly.

Next the public must be made to understand that the telephone network is a single enormous distributed computer that needs centralized coordination. It can't be run by a number of groups with conflicting interests.

There must be an acceptance of John Stuart Mill's "greatest good for the greatest number" goal. Some services must necessarily subsidize others to achieve universal service. In the long run, it benefits everyone.

And lastly, the disagreeable fact has to be faced that the United States has let other countries surge ahead of it in the telecommunications field. This country is years behind where it could have been technologically. It must regain its leadership in technology if it is to flourish.

As Congressman Al Swift (D-Washington) of the House Telecommunications subcommittee sees it, "It depends on whether you can get the public to care. In the past everyone yawned."

The country has got to stop yawning. Voters must be galvanized and make their public officials aware of what needs to be done. They must start backing candidates—from the president on down—who pledge to act and rectify the past mistakes and blunders in the telecommunications field. Mike Wallace on "60 Minutes" should be exposing the millions and billions wasted. *The New York Times* should be having editorial apoplexy.

But that's only the first step.

Then Congress must finally get off its collective rear end and act. But before cranking up the legislative machine, bipartisan hearings should be held so that the views of persons and companies with a stake in the matter may be heard.

The issues of what lies ahead for this country in telecommunications might not be as "sexy" as Fawn Hall, Oliver North, and the Iranscam investigation, but in the long haul, they are infinitely more important. Testimony should be heard from present and former telephone executives and engineers, equipment manufacturers, competitors in the long distance market, present and former FCC and PUC members, and from independent experts in communication technology. Get it on the record.

National defense and welfare issues should be addressed in terms of the ability of our telecommunications system to survive natural and man-made disasters. The hearings should examine the wasteful duplication which continues to take place, and realize that the talents and resources that are consumed in contrived competition could instead be used to create a modern and well-integrated system.

The question of government ownership will undoubtedly be raised. In most other countries, communications systems are operated as departments of government or as public corporations which are tightly controlled or even financed by the government. Right away, any such arrangement should be rejected for the United States. Why would anyone suppose that a government-owned telecommunications network would run any better than the post office?

The framework of the American government makes such a

set-up ill-advised. As a government agency, telecommunications would have to fight continually for appropriations, and
would more than likely find itself far down on the list of priorities. Defense, welfare, health care, farm subsidies, and other
Congressional concerns with well-established lobbies and
widespread public support would shoulder out telecommunications needs. Not knowing from year to year what kind of
money to expect is not conducive to long-term planning.

Beyond these points, the government itself should be disin-
clined to allow such an arrangement since it would lose a
very considerable amount of money if deprived of the income, sales, and excise tax revenues it now receives from the
industry.

Even though the government should not take over
telecommunications, some sort of strong central control is
mandatory. Such leadership is needed to direct research, enforce standards, develop pricing philosophies, monitor service quality, coordinate planning, and negotiate from a position of strength with other telecommunications authorities
throughout the world.

In other words, something on the order of—that's right—
the old AT&T.

Once the hearings are over, Congress can stop neglecting
its responsibilities and get down to the crucial business of
legislating.

Over the last twenty years Congress has been content to
stay on the sidelines while the FCC and the federal courts
fashioned their ill-conceived and poorly sewn patchwork
quilt of telecommunications policy. In fact, it has done nothing since passing the Telecommunications Act of 1934. The
world isn't the same place it was fifty years ago. A lot has happened since then, but the word apparently has failed to reach
Capitol Hill.

Not that there haven't been some attempts. In 1976, AT&T
backed legislation that would have restored the provisions of
the 1934 act and also affirmed the telephone network as a
natural monopoly with universal service a national goal. The
proposal went nowhere.

Congressman Ed Markey (D-Mass.) said recently in a

speech to a Telephone Association seminar, "It's now time to give your pivotal and radically evolving industry direction and stability, and consumers protection and confidence. I believe that begins by resurrecting the FCC as a regulatory agency. It's time to reacquaint the FCC with its statutory obligations to protect the public interest."

In 1986, Senator Robert Dole (R-Kan.) proposed a bill that would have restored full regulatory powers to the FCC. It would have also spelled out which areas of telecommunications would be open to competition and which would be regulated.

But Dole's proposal followed earlier ones into oblivion.

It's time Congress faced up to the consequences of its indifference and passivity. Had it taken some sort of timely action, it could have prevented divestiture or at least softened its impact. It's going to be much more difficult undoing the harm of divestiture in future legislation than it would have been to stop it in the first place. In any case, Congress must now establish a direction for the nation's telecommunications and its regulation.

And what should new legislation do?

First, it should reconstitute the FCC as the regulating agency for interstate—not intrastate—telecommunications and get Judge Greene and the Justice Department out of the industry. The regulatory authority usurped by the FCC must be restored to the states. On top of this, Congress should once and for all acknowledge that telecommunications is a highly technical business and not well served by amateurs. Any legislation should give technically qualified people a greater regulatory role and concomitantly reduce the domination of lawyers. Superior regulation is essential with personnel equipped technically to join in the effort to regain world leadership.

In short, the challenge is to provide a regulatory body empowered to furnish the world's best telecommunications service at the lowest prices—which could well be 50 percent lower than what they are now.

There is a small ray of hope in the fact that FCC chairman Dennis Patrick went on record as recognizing that his agency is a creature of Congress and subject to Congressional overviews and the FCC's goal should be the consumer's welfare. Time will tell.

The second thing on the legislative agenda should be to reestablish the network as a unified and integrated entity. AT&T, over perhaps a five-year period, should buy up the surviving long distance common carriers. It's not a far-fetched idea. It's more than likely that in time these competing companies would quite happily sell off their investments since many are not doing well. They have been finding that the network is indeed a natural monopoly and where AT&T has been allowed to compete on equal terms, the other common carriers have had trouble staying alive.

Next, the legislation should require that standards be set and maintained for the entire network. Any equipment to be plugged into the network would have to meet established standards. It can't be emphasized enough that standards are essential in maintaining communications compatibility. It is impossible to maintain a communications network without compatibility.

Let it be clear that it's not necessary for AT&T to own all the equipment as it once did. No, but what is vital is that the design, manufacture, and maintenance of equipment meet established safety and performance standards. Then the man in the moon can own the equipment, if need be. Even the FCC's own study, conducted by the National Academy of Sciences, recommended that.

Finally, Congress should recognize consumer needs and protect them. A customer should be able to place a single call to order or repair anything. The Baby Bells should be required to give the customer the option of end-to-end maintenance. One call is all it should take. Of course, with the customer owning equipment now, such service can't be made universal or mandatory. But at least the customer should have the choice.

Ma Bell is dead and she isn't going to be brought back to life. That would be a political impossibility and an administrative nightmare. At the very least, the top executives of the Baby Bells, who have fared so well financially, would do their utmost to block such an attempt.

Whether it is for the best or not, the era of a one-owner nationwide telephone system is over. Instead there are AT&T, seven Baby Bells, the more than twelve hundred independent companies, the MCIs and all the others, each concerned only with its particular niche in the network.

But the need for central direction is urgent, and Bellcore, as it stands today, isn't going to do the job. It has too many masters and not enough power. The situation has to change, and the best solution would be for AT&T to absorb Bellcore and strengthen it. The Baby Bells should be required to support Bellcore financially, much as license contract fees were once used to support similar efforts in the Bell System. A revitalized Bellcore, as a unit of AT&T, should then be given clearly defined authority to enforce standards and give advice where desired.

With authority over standards and service in a reconstituted Bellcore, there would be no need for AT&T to once again own the local operating companies. In fact, to endrun any future antitrust action, it would be well for AT&T not to own the local companies lock, stock, and telephone wires. But there doesn't seem to be any reason to prohibit a percentage of the stock in AT&T's hands.

Once AT&T is given responsibility for the nationwide computer of telecommunications, it can reinstate the service quality measurements and reports of the Bell System days. Judge Greene, the FCC, and the Justice Department have been too busy carving up the system to pay much attention to such trivialities. AT&T could bring back the equivalent of the Green Book and once again start measuring and reporting on service quality and enforcing changes where needed. The FCC should be responsible for a general overview of these matters. Once quality standards are reestablished, they can be reevaluated and improvements made.

AT&T should also study in depth the public's need for new services which would increase the ability to communicate more effectively. A starting point might be to select one thousand individuals representing a broad spectrum of the population, industry, and government, and have their daily activities and interactions with others observed over a period of several weeks. Data could be collected on all time—productive or otherwise—spent communicating.

Then the study group could be offered new services and the effects of those services studied in terms of time savings, communications effectiveness, productivity, and so forth. The services found to be profitable and beneficial could then be offered to the public as a whole.

There should also be studies regarding pricing of services. Twenty-five years ago, the trend was toward flat-rate pricing. Indeed, a number of Bell System managers supported the concept of uniform charge—possibly twenty-five cents—for all three-minute, directly dialed calls no matter the distance. In other words, a call from New York City to Newark, New Jersey, would cost the same as a call from New York to Seattle.

This made certain economic sense. Transmission costs were dropping and by simplifying long distance accounting and billing, there wouldn't be much difference in the cost of providing a long distance and a local call. Also, at the time, local unlimited calling areas were being expanded to include entire metropolitan areas.

More recently the trend has moved to cost-basing, as would be expected in a more competitive environment. This has led to vast expenditures for such things as local measured service, which provides detailed recording of data on local calls. Under this system, on local calls a customer is charged not only by distance but also according to the duration of the call and the time of day. In Pennsylvania alone more than $30 million was spent in 1983 to install equipment to do this measuring.

The direction to take in pricing is probably a mixture. On the one hand, sinking an enormous amount of money into equipment that does nothing but record detailed charges isn't justified, but on the other some cost-based pricing is necessary to prevent abuse.

A perfect example of abuse is the once free directory-assistance service. Anyone wanting the number for the local library called directory assistance and got the information without charge. The cost of providing this service (the operators' salaries and equipment) was distributed among all telephone users. But studies showed a few users were making a disproportionate number of calls. In this case, the present system of charging for these calls is fairer for everyone.

Another area where cost-basing makes sense is with non-voice network users. Someone plugging a computer modem, for example, into the network may tie up costly transmission and switching facilities for hours. It's only right that he should pay on a timed and not a flat-rate basis.

When measuring of a service can be done at a reasonable cost, then cost-based pricing is in order. But if there is a significant savings in flat-rate pricing, then that should be the method used even if some cross-subsidization results.

There is one issue that should not be lost in the pricing shuffle. And that is that some basic form of communication—even if it has to be subsidized—must be made available to everyone. Theodore Vail's goal of universal service should be maintained even if some compromises have to be made.

There's a difficult time ahead. Playing catch-up is always hard.

The changes wrought by the FCC and the courts have encouraged AT&T and the Baby Bells to diversify into other lines of business. There is too much needing to be done in telecommunications for attention to be diverted into stock brokering and real estate. The country is already twenty years behind where it should be in providing modern services. And billions of dollars each year are being squandered needlessly.

It's a pity, a waste, a crime. And it's got to be stopped. We must catch up, and we must stop this wasteful spending of the public's money. The public has no way of even knowing what it needs and wants in its telecommunications services, since it doesn't know what the possibilities and the choices are. The public must be told.

Without prodding, government officials will be content with the status quo. And don't expect managers who now have cushy, high paying jobs with the Baby Bells to raise their voices.

Still, change is possible. The public, disgusted with deteriorating telephone service, has got to take a cue from Peter Finch in the movie *Network*. It's got to throw open windows and shout from the rooftops, "We're mad as hell and we aren't going to take it any more."

Otherwise, silence becomes complicity.

Common sense must be returned to telecommunications. The job ahead is so vast, so multifaceted, so complex that it will take the combined efforts of the integrated industry at least ten years to restore the United States to leadership.

The problem now is that no one is in charge. Congress must remedy this. It must recognize the total failure by state and federal regulators to serve the best interests of our citizens. Then it must revamp the FCC and make it a strong overseer that will encourage the creation of a unified system out of many and diverse parts.

The time to get started is now.

Today.

19

How Does It Work?

To the average person, it's very simple.

Telling his best friend what's happening in a heart-to-heart telephone conversation. Calling his doctor to ask about a medicine's side effects. Making a toll-free call to order throw pillows from the Spiegel catalog.

For the caller, the hardest part of a telephone call is often what to say. But lurking between the movement of picking up the handset and finally replacing it in the cradle, thousands of separate, complex, intricate, and, for the layman, near-miraculous operations take place. A degree in engineering is needed to understand all the nuances, science, and technology of the modern telecommunications system. But even without one, some of the basics are within reach.

So here, then, is a Cook's tour of the technology.

Start with the human voice. It's a sound that rises and falls, has different intensities (or amplitudes) and pitches (frequency). The goal is to transmit the "information" in the voice to the desired destination.

There are two ways information can be transmitted by electrical means—by analog or digital signals.

Analog signals, used in telephony right from the days of Bell and Watson, very much resemble the pattern of the voice, using a continuous range of amplitudes and frequencies. Humans can generally hear within a frequency range of 30 hertz—that's the unit of measurement for frequency—to 20,000. But to transmit intelligible conversation, a range of only 200 to 3500 hertz is needed.

Digital signals, on the other hand, are more like a code sent in a stream of on/off pulses. The information in the transmission is found in the configuration of the pulses. It is cheaper to transmit digitally, which is why there is a shift today from analog methods to digital.

Whichever method is used, there still remain three steps in getting the "Come here, Mr. Watson. I want you." from one telephone to the other. First, the speaker's sound waves have to be translated into electrical signals, then these signals have to be conveyed to their intended destination, and third, the electrical signals must be reconverted into sound waves.

Of course, that's an oversimplification of the whole process, and in the not-so distant past, there were all sorts of problems, especially in the middle, or transmission step. Some of these included the signal losing too much power (attenuation) over the distance traveled. Or the sound was distorted or interfered with. Or signals were bouncing back to create an echo. And foreign tones were "singing" on the line making it hard to hear.

Scientists and engineers labored long and hard to develop means for keeping these problems within acceptable limits.

But today, these problems are largely avoided by using a technique called pulse code modulation, or PCM. With it the original voice signal is encoded into a series of numbers. The numbers then get transmitted. It doesn't matter if they are distorted as long as they remain recognizable. The numbers are then used to rebuild the voice signal at the other end.

The encoding is accomplished by "sampling" or measuring the voice signal amplitude at a given moment. The amplitudes are assigned a number—usually from zero to 255. Within one second, 8,000 measurements of amplitude are taken. This produces a series of numbers that, when put back together in the same sequence, will reproduce the original sound.

These numbers are then converted into an eight-digit binary number for transmission. Since any binary number—no matter how long—can be represented by a series of ones and zeroes, the number can be translated into a sequence of on/off electrical pulses—such as "off" for a zero and "on" for one. So the original voice signal can then be expressed in terms of 64,000 pulses—that's the eight digits multiplied by 8,000 samples—each sound.

While that may seem like a lot of pulses, they are extremely short and that number doesn't come close to the transmission medium's capability. Therefore the pulses of several conversations can be combined—or interleaved. Typically 24 speech channels are combined to form a signal of 1.54 million pulses per second—or a 1.54 megabit signal. This 1.54 megabit signal—a T1 channel—is a basic building block in digital telecommunications.

The introduction of coaxial cable, microwave, and optical fiber waveguide transmission has greatly increased the speed or bit rates at which transmission is possible. Now a number of these T1 groups can be assembled to create a system combining hundreds, even thousands, of voice channels into a single long and very rapid series of binary pulses. This combination process is called "multiplexing."

With this system, distortion and interference are eliminated as long as the presense or absence of the on and off pulses can be detected. To insure that, intermediate regenerator points (at approximately every mile with wire transmission, and ten to twenty with optical fiber) are established. At these points electrical circuits recognize the incoming pulses then generate corresponding new pulses to be sent on. This process is called regeneration and differs from conventional amplification which simply increases the power level of the incoming signal, distortion and all.

At the receiving end, the process is reversed. The interleaved signals are separated into individual speech channels, each with a string of pulses at a 64,000 per second rate. These are decoded into signals whose amplitudes are determined by the 0 to 255 binary code levels. The resulting 8,000 pulses per second of varying amplitude are then passed through an electrical filter that regenerates an exact replica of the original voice signal.

That's how sound gets from one point to another. But such transmission would be useless if the sound didn't get to the right destination. That's where switching comes in.

Telephone switching is not unlike railroad switching, where the proper setting of track switches assures each train is directed to its proper destination. In telephony, calls are directed through a series of switches that are determined first by the number the customer dials and then by the logic built into the switching machine. Modern switching systems are electronic and the electronic instructions are in the form of a stored computer program—software—which can be easily modified.

There are two types of switching systems in use. The older system is known as "space division" switching and it provides each call with a physically separate path. The newer system is called "time division." Here calls may share the same physical pathway once they are given a digitally coded time slot.

The difference between space and time division switching is the difference between using ten streets to get ten cars from one side of the city to another or sending ten cars in sequence over just one street. To carry this analogy further, in the latter case, the ability to handle traffic is greatly increased, but in order to prevent collisions, much more complicated controls are needed.

Time division is being used for almost all new switching installations and has the advantages of lower costs and better reliability and quality.

With either space or time division, there must be something directing the operation—"throwing" the proper switches. That something is a specialized digital computer, run by an extremely complicated software program.

Let's look at one part of this computer operation. A unit called a scanner examines the lines and trunks in a central office in a continuous rapid cycle. If the condition of a line has not changed—whether it had been idle or in use—from the last cycle, no action is needed. However, should there be a change, the scanner sends the information to central control so proper action can be taken.

Say the scanner finds a line going off-hook—someone has picked up the handset to make a call—the scanner will refer that information to central control. Control assigns a unit called a register to the line to store the digits as they are dialed. The control will also have a dial tone sent out on the line to the customer, telling him it's ready for his instructions. It will also arrange for more frequent scans on this line, anticipating dialing action.

All this is done faster than a user can raise the handset to his ear.

Besides the information from the scanner, central control must deal with information in the register—the number the customer wishes to reach. Central control dips into its vast memory banks and determines where the number is, which path is needed to reach it, pertinent billing information and special instructions such as call waiting, call forwarding, and so on. Finally the necessary switches must be closed to establish the connection. These may be actual physical switches or they might be assigned time slots.

Although central control can deal with only one simple operation at a time, it finishes this so quickly—in about a millionth of a second—that thousands of customers get the equivalent of fulltime attention although they are actually being dealt with sequentially.

It should be noted that the trend is moving rapidly toward pulse transmission and time division switching. This means that soon calls will be both transmitted and switched in the form of strings of binary digital pulses. The traditionally separate telephone disciplines of transmission and switching will have then come together and the whole system will be analogous to a digital computer.

Then, keeping in mind that not only voice messages but video and data as well can be transmitted on digital basis, something very exciting is on the horizon.

The Integrated Services Digital Network or ISDN.

Picture this. With ISDN, a customer could make ordinary telephone calls, send facsimiles, have access to videotext, select the TV movie he wants to watch, have his gas meter read, get stock market reports, order merchandise from a catalog, and on and on, all simultaneously.

It will be a revolution in telecommunications and a revolution in the way we live and do business.

And it will be accomplished, in part, by a new transmission concept called packet switching.

Here a stream of binary pulses is broken into segments—packets—usually of some 1,000 bits long. These packets, each with its destination address, sequence order, and other information, are frequently called envelopes. They are sent individually to their destination over the most efficient route available at the moment. The route can differ from one packet to the next. The packets of data are stored briefly at nodes, and forwarded at high speed to the next node when a suitable path is available.

At the final destination, the packets are reassembled in proper order to form an identical copy of the original pulse stream.

By collecting and storing data before transmission, high speed transmission facilities can be used more efficiently in sending data that originated at lower speeds. In addition, there are nearly unlimited alternative routing paths available during busy traffic conditions or circuit failures.

Underlying these basics of modern telecommunications technology described here is a collection of principles often called Information Theory. Claude Shannon, a former Bell Telephone Laboratories scientist, is responsible for its mathematical formulation.

What the theory does is give a quantitative measurement of the maximum amount of information that can be transmitted without error over a given channel at a given time. Just as Einstein set the speed of light as the speed limit of the universe, Shannon proclaimed the limits of our ability to transmit information. While the rate can never be exceeded, it is the engineer's challenge to get as close to the limit as is economically possible.

Statistics are the heart of Information Theory. By knowing the type of information to be transmitted, efficiency can be improved. Take the English language. Some letters are used more frequently than others. Some letters bunch together more frequently. And, then too, given enough letters of a word, it's usually easy to guess what will follow. (This is the basis of that oh-so successful TV show, "Wheel of Fortune.") So, given the letters R-O-N-A-L-D R-E-A-G-, with a high degree of certainty it can be said the next two letters are "A" and "N" and they don't have to be sent.

Video transmission, which is more than 99 percent redundant, is made up of individual picture elements called pixels or dots. If the picture is of a grape, the dots will be purple in the grape and will only change at the edges. So in transmitting the information on the grape, it should be possible to only send signals when the pixel color changes. This would greatly reduce the rate of information transmission needed because less information would need to be sent. These and other techniques suggested by Information Theory are being investigated to make our transmission systems more efficient.

Predicting advances in technology is a thankless task. Think of that poor Royal Society physicist who stated flatly, in 1895, "Heavier than air flying machines are impossible." Or Robert Milliken, a Nobel laureate in physics, who in 1923 stated, "There is no likelihood man can ever tap the power of the atom."

Forecasting bloopers could fill volumes. But undaunted by past records, herein are some predictions of what's in store for telecommunications.

First off, it is quite safe to predict that digital transmission methods will almost completely replace analog systems within a short time. Also, digital telephones will soon be available. With these, conversion to binary coded pulses will take place at the point of origin—that is the telephone set itself—and not in the central office.

Optical fiber will be used almost exclusively as the physical medium for transmission of calls, although copper wire and microwave radio will continue to be used for some years.

In switching, new installations will be digital time division systems rather than space division. Systems with distributed control are more likely to be chosen over those with centralized control. And the next generation of switching machines may use optical instead of electronic switching, although this is still in the early research stages.

In terms of the information transmitted, more sophisticated techniques to get closer to Shannon's theoretical limits will be developed, allowing information to be sent more efficiently.

There are incredible, dramatic advances to come. No matter what the politicians, lawyers, and bureaucrats do to obstruct progress, scientists and engineers will continue to enhance the technical capabilities of our telecommunications system.

It's their nature. Otherwise they might have listened to Charles Duell, director of the U.S. Patent Office, when he declared in 1899 that "Everything that can be invented has been invented."

Glossary

A Journey Through the Jargon

ACCESS CHARGE—Either 1) a charge by the local Bell companies to intercity carriers (AT&T, MCI, etc.) for connection to the local networks, or 2) a direct charge by the local Bell companies to customers, in part offsetting the subsidy formerly provided from long distance revenues.

AMPLIFICATION—The process of strengthening a communications signal that has been attenuated—weakened—in its transmission path. In amplification, unlike regeneration, both the desired signal and undesired distortion or interference are strengthened. See ATTENUATION and REGENERATION.

AMPLITUDE—The voltage, intensity, or strength of an electrical signal representing voice, video, data, etc.

ANALOG—A signal employing a continuous range of amplitudes and frequencies. Generally the electrical signal bears a direct resemblance to the nonelectrical (acoustical or video) input. See DIGITAL.

ANI (Automatic Number Identification)—The process or equipment used for determining and recording—usually for billing purposes—the number of a calling customer.

ATTENUATION—The loss in intensity or strength of an electrical signal as it is transmitted over distance.

AUTOMATIC NUMBER IDENTIFICATION—See ANI.

BANDWIDTH—The range of frequencies that can be transmitted over a communications channel. Bandwidth bears a direct relationship to the quantity of information that can be transmitted within a given time. See CHANNEL CAPACITY.

BELLCORE (Bell Communications Research Corporation)—The organization formed at divestiture to serve as a central point for coordinating government communications needs, and also as a centralized staff organization for the Bell operating companies. Bellcore is owned in equal shares by the seven Baby Bells.

BELL SYSTEM—The term for the total organization controlled or associated with AT&T prior to divestiture. This included AT&T headquarters, Bell Telephone Laboratories, Western Electric, and the Bell operating companies. The Bell System ceased to exist as of January 1, 1984.

BINARY—A numbering system or electrical communications system based on two and only two circuit states—for example, on/off or 1/0. This is in contrast to the decimal numbering system, which has ten values (0-9).

BIT—A contraction of "binary" and "digit" meaning just that. In the binary numbering system, each digit (1 or 0) is a bit. A bit can also be regarded as the smallest unit of information stored in a computer, much as the electron is the smallest unit of matter.

BYTE—A grouping of bits, i.e., a word, typically eight, sixteen, or thirty-two. A byte is to a bit what a word is to a letter.

CARTERFONE—A device to connect private radio systems to the telephone network. It was invented and marketed by Carter Electronic Corporation of Texas in 1968. AT&T would not allow connection until the FCC, in its Carterfone decision, ordered the connection. The decision was the first significant break in the Bell System's long-standing monopoly in providing terminal equipment.

CATHODE RAY TUBE—See CRT.

CCIS (Common Channel Interoffice Signaling)—A system which separates a voice channel from its associated signaling and supervision, allowing more direct and rapid routing. This system determines availability of a suitable routing path before the connection is actually set up and reserves the necessary parts of the circuit. The introduction of CCIS reduced the setup time for long distance calls from ten seconds to two and reduced the opportunity for fraud.

CELLULAR MOBILE RADIO SERVICE—In contrast to older mobile systems, it has almost unlimited call-carrying capacity. Because of the short transmission range at the high frequencies and the lower power used, channels can be reused within a fairly short distance. The key to the operation of the cellular system is that conversations are handed off between base stations as the vehicle moves from cell (a small geographical area) to cell. This is accomplished with a high-speed electronic switching machine. The cellular mobile telephone was tested successfully in Chicago in the early 1970s, but its introduction was delayed more than a decade by the FCC.

CENTRAL OFFICE—A switching machine location where customers' lines are connected to each other or to interoffice trunks. The terms central office, wire center, switcher, and node are used interchangeably.

CENTREX—An arrangement for business customers that provides, in a telephone central office, the same services normally furnished by a private switching system (PBX). Because each station line must be connected individually to the central office, this service is economically viable only when the customer's location is relatively near the central office. See PBX.

CHANNEL—A path for carrying information in electrical form.

CHANNEL CAPACITY—The quantity of information that can be carried in a specified time on a channel. The capacity is usually expressed as bandwidth (in hertz) or in bit rate (number of binary digits—bits—per second). See BANDWIDTH and BITS.

CHIP—A very small sliver of silicon onto which complex circuitry—including transistors, resistors, diodes, capacitors—have been etched or otherwise applied. This technology permits a great many circuit functions to be contained in a very small space. See LSI and VSLI.

COAXIAL CABLE—A cable structure using two conductors,

one of which is hollow and completely surrounds the other. Transmission takes place at relatively high frequencies, upon which a large number of communications channels are multiplexed. See MULTIPLEXING.

COMMON CHANNEL INTEROFFICE SIGNALING—See CCIS.

COMMON CONTROL—A method of operation used by all modern electro-mechanical and electronic central offices, whereby most logical and memory functions are centralized in units of equipment similar to those used in computers and called—depending on the type of equipment—markers, senders, common control units, signal processors, etc.

COMPUTER I, II, III—A series of inquiries and decisions by the FCC, from 1966 to 1986, aimed at clarifying the boundaries between data transmission and data processing.

CONSENT DECREE—A legal agreement between contestants that generally ends litigation in return for an agreement on certain future behavior. For example, the antitrust suit brought in 1949 by the Department of Justice against AT&T was terminated in 1956 by a consent decree limiting the types of business in which AT&T could engage.

COST-BASED PRICING—A philosophy of setting prices for a service based on the actual cost of providing it, with no averaging across services or cross-subsidization.

CPE (Customer Premises Equipment)—Equipment located on the customer's site that may or may not be owned by the customer. See TERMINAL EQUIPMENT and STATION EQUIPMENT.

CROSSBAR—A type of switch widely used in telephone central offices from the 1940s into the 1980s. It consists of a grid of horizontal and vertical contacts which are selected by the operation of the appropriate horizontal and vertical file magnets. The physical structure, with bars across the face of the switch performing the physical closures, gave the device its name. A typical crossbar switch might be 30 inches wide and 12 inches high. This is now obsolete technology.

CROSSTALK—The interference of unwanted signals with the desired one, usually by electrical induction in the transmission path or by improper switching action in a central office.

CRT (Cathode Ray Tube)—A piece of equipment using an electron beam to trace an image on a phosphorescent screen. Widely used in television, computer terminals, and test equipment.

CUSTOMER PREMISES EQUIPMENT—See CPE.

DATA—In telecommunications, information stored or transmitted, usually in binary digital form.

DDD (Direct Distance Dialing)—A system that gave every customer a unique ten-digit number and allowed calls to be dialed without going through an operator. Introduced by the Bell System beginning in 1951, it allowed much faster long-distance communications.

DEMODULATOR—A device separating a carrier frequency and a communications signal. See MODULATOR.

DIGITAL—A device or system operating in a mode that expresses information numerically (usually in binary form). As long as the number representing the information is preserved, the original information can be reconstructed perfectly, despite intervening distortion. See ANALOG.

DIRECT DISTANCE DIALING—See DDD.

DISTORTION—The undesired alteration of the amplitude or frequency of an electrical signal as a result of imperfect equipment or outside interference. A serious problem in analog transmission but not much of one in digital since these signals can be regenerated. See REGENERATION.

DISTRIBUTED PROCESSING—A recent development in which many of the intelligent electronic switching functions are divided among various units of equipment rather than being concentrated in a single central unit. Used, for example, in AT&T's #5 ESS.

DNHR (Dynamic Non-Hierarchical Routing)—A new concept for interconnecting nodes in the network without using the traditional hierarchy of Primary Center, Sectional Center, etc. The maximum number of links is reduced, and the speed of call connection cut to less than 2 seconds.

ECHO—The reflection of signals back to their source caused by imperfections in telephone equipment. Because the returning signals are slightly delayed, they interfere with the outgoing signal. The longer the delay and the stronger the reflected signal, the more annoying the echo is to the speaker.

ECHO SUPPRESSION; ECHO CANCELLATION—Techniques to minimize the adverse effects of echoes.

ELECTROMECHANICAL—The operation in central office switchers and PBXs that use electrical/mechanical devices, such as relays, step-by-step, and crossbar switches. This type of equipment has largely been phased out and replaced with electronic switching systems. See ESS.

ELECTRONIC SWITCHING SYSTEM—See ESS.

END OFFICE—A local central office at the lowest level in the switching hierarchy, connecting directly to customers' lines.

EQUAL ACCESS—A system permitting all other common carriers (such as MCI) the same quality of interconnection to the Bell companies' local networks as AT&T.

ESS (Electronic Switching System)—Systems with few if any moving parts, using electronic logic and memory and making connections using solid state electronics rather than mechanical contacts. All modern systems are of this type, and have almost completely replaced the older electromechanical ones. See ELECTROMECHANICAL.

EXCHANGE—A local area with uniform telephone rates. Frequently this is the same as the area served by a single central office code—the first three digits of a telephone number—but not always. Especially in metropolitan areas, there is more than one central office code within an exchange area. The term was also used in the modification of final judgment in an entirely different sense to designate a much larger area. See LATA.

FCC (Federal Communications Commission)—An agency of government chartered by the United States Congress in 1934 to regulate interstate common carrier and broadcast communications.

FILTER—An electrical device that suppresses the transmission of certain signal frequencies, thus permitting separation of frequencies. Used in, among many other things, modulation/demodulation. See DEMODULATOR and MODULATOR.

FINAL ROUTING—The path between two points taken after one or more shorter, less costly paths have been tried. See HIGH USAGE ROUTING.

FREQUENCY—The number of reversals per second in signal polarity. With audio signals, frequency is detected by the human ear in terms of pitch—a high-pitched sound is at a higher frequency than a low pitched one. Human hearing extends to a frequency of about 20,000 cycles per second (or hertz). Telephone channels generally extend to about 3,500 hertz, which is adequate for intelligible speech but not high-quality transmission of music. See HERTZ.

GROUND RETURN—A type of telephone circuit using only a single wire conductor and depending on the conductivity of the earth as a return path. An unsatisfactory system subject to interference, it was phased out at the turn of the century.

HERTZ—A unit of frequency measurement equaling one cycle per second. It is often prefixed with kilo- (thousand), mega- (million), or giga- (billion).

HIGH USAGE ROUTING—A fairly direct—and low cost—telephone traffic routing between originating and terminating points, with a quantity of trunks engineered to handle about 80 percent of offered traffic. The remainder, or overflow, is sent on less direct high-usage routing or to a final route. See FINAL ROUTING.

IMPEDANCE—A characteristic of a circuit that determines how a communications signal will be affected when it is applied to the circuit. Impedance varies with the frequency of the signal and is measured in ohms.

IMPEDANCE BALANCE—The matching of impedances of connected circuits to maximize energy transfer and minimize echo and distortion.

INFORMATION THEORY—A body of knowledge dealing with the information content of a message, allowing efficient means of encoding and transmitting of that message.

ISDN (Integrated Services Digital Network)—A concept taking advantage of the uniform nature of signals in digital format so that various types of information can be transmitted over a single network.

KEY TELEPHONE—A system used largely by small business customers, with each phone having access to a number of lines by pressing buttons on the telephone set.

LASER (Light Amplifications by Stimulated Emission of Radiation)—A device which generates coherent (uniform frequency) light beams electronically. These can be modulated and the resulting signal transmitted as a light beam through optical fiber. See OPTICAL FIBER.

LAISSEZ-FAIRE—An economic theory popular during the last century that espouses unfettered competition as the solution to most economic problems. It is generally out of favor today.

LARGE SCALE INTEGRATION—See LSI.

LATA (Local Access and Transport Area)—Under the provisions of the Bell System divestiture, telephone service areas were divided into approximately 160 "exchanges." Service within them became the province of the Baby Bells on a regulated monopoly basis. The interexchange services were to be unregulated and went to AT&T and the other common carriers. To avoid confusion with the historical definition of "exchange," the new serving areas were renamed LATAs. They are sometimes referred to as "service areas."

LED (Light Emitting Diode)—A device for converting electrical energy to light. In optical fiber transmission, it can be used as an alternative to the LASER. It is also used in illuminated displays, such as calculators and digital clocks. See LASER and OPTICAL FIBER.

LIGHT AMPLIFICATION BY STIMULATED EMISSION OF RADIATION—See LASER.

LINE SWITCHING—A system commonly used to make end-to-end communications circuit connections. See MESSAGE SWITCHING.

LOADING COIL—On voice frequency circuits, an inductor connected to the line at specified intervals, usually six thousand feet. This changes the impedance of the circuit so that higher frequencies are cut off, but the lower frequencies of voice transmission are enhanced. In the early days of telephony, loading made efficient long-distance communications possible. Today voice frequency transmission is rapidly being replaced by digital transmission.

LOCAL ACCESS AND TRANSPORT AREA—See LATA.

LOGIC—In telecommunications, operations involving logical decisions, such as "and" and "or" functions. For example, a circuit that will deliver an output signal if and only if a signal is present on each one of several input terminals is one type of logic circuit, called an "and gate." Logic functions, which often get quite complex, are indispensable to circuit switching. In older electromechanical central offices, logic was built into the equipment and could be changed only by physically rearranging wires. In electronic switching, much of the logic is stored in computer programs and can be modified by typing in program changes.

LSI (Large Scale Integration)—The technology of combining a great many circuit elements (diodes, transistors, resistors, etc.) on a single silicon chip. The resultant small size, low cost, low power requirement, and mass duplicability of the circuits have made the modern computer and telephone switching system possible.

MASER (Microwave Amplification by Stimulated Emission of Radiation)—A predecessor to the LASER, this operates on lower frequencies and has been practically replaced by the newer technology. See LASER.

MCI (Microwave Communications Incorporated)—The first common carrier company to compete with AT&T, it was founded in 1969.

MEMORY—In computers and electronic switching machines, the capability of storing (remembering) information for later retrieval. A simple example is electronic storage in the central office of each customer's class of service including his rate plan and options chosen such as touchtone and custom calling. Memory may exist in various physical forms, but magnetic memory storage is the most common.

MESSAGE SWITCHING—An arrangement where information is transmitted between an originator and a receiver by sending a message to an intermediate point rather than over a built-up connection as in line switching. The information is stored at the intermediate point and advanced towards its destination when a path is available. It may be stored and forwarded several times before reaching its destination. The mail and some telegraph systems have long used message switching. The packet switching systems just coming into use now also employ this method. Formerly referred to as "Store and Forward." See PACKET SWITCHING.

MFJ (Modification of Final Judgment)—The 1982 document that terminated the Department of Justice's antitrust action begun in 1974. It was called a "modification" because it amended the 1956 Consent Decree—which in turn terminated a 1949 antitrust action.

MICROWAVE—Electromagnetic radiation above a frequency of approximately 2 gigahertz. Microwave radio transmission has been used widely in telecommunications since 1948. Its use is declining now as optical fiber transmission becomes more common. See OPTICAL FIBER.

MICROWAVE AMPLIFICATION BY STIMULATED EMISSION OF RADIATION—See MASER.

MICROWAVE COMMUNICATIONS INCORPORATED—See MCI.

MICROWAVE RELAY—Microwave radio transmissions travel in straight lines and can only be sent over line-of-sight distances of about 15 to 30 miles, depending on the terrain. At these intervals, the signals are received, amplified, and retransmitted. The transmitting and receiving antennae are usually mounted on towers up to several hundred feet high and are a common sight on hilltops. The collection of towers, transmitters, and receivers is known as a microwave radio relay system.

MODEM—A unit of equipment in which the functions of the modulator—at the transmitting end of a circuit—and demodulator—at the receiving end—are combined. The word is an acronym of modulator and demodulator.

MODIFICATION OF FINAL JUDGMENT—See MFJ.

MODULATOR—A device combining a signal, such as a telephone voice signal, with a carrier frequency for transmission. The carrier frequency, usually much higher than the signal, is easier to transmit over long distances and also permits a number of voice signals to be combined on it. See DEMODULATOR.

MULTIPLEXING—The process of combining a number of signals for efficient transmission. A good example is T1 carrier, which provides 24 voice signals to be interleaved—or multiplexed—for transmission over a single path. Multiplexing may be analog or digital, and may also be performed in multiple stages. A number of T1 systems may in turn be multiplexed to create a higher level system, such as T3.

NCC (National Coordinating Center)—A group established

to provide rapid contact between government agencies and telecommunications carriers during emergencies.

NECA (National Exchange Carrier Association)—A group formed after divestiture to, among other things, oversee a pool of funds supplied by interexchange carriers and distributed to local exchange companies as subsidies.

NEGATIVE FEEDBACK—A concept developed in 1927 that made modern telecommunications possible. By feeding back a small portion of the output signal to the input in inverted (negative) form, distortion of the signal is greatly reduced and high quality amplification becomes possible.

NODE—A point where telephone lines or trunks are brought together for interconnection. See CENTRAL OFFICE.

OPTICAL FIBER—A very thin—much smaller than human hair—and transparent glass thread capable of transmitting light over a great distance. Since what is transmitted is a light wave, the term "optical fiber waveguide" is often used. The light wave is modulated before transmission by many telecommunications signals, thus furnishing a very efficient method of long distance communications. Optical fiber transmission is rapidly replacing most other methods. Since no electricity is transmitted, electrical interference is not a problem.

OTHER COMMON CARRIER—A term for interoffice telecommunications service providers other than AT&T. See MCI.

PACKET SWITCHING—A relatively new method of transmitting data. Messages are broken into relatively short bursts—or packets of data—and transmitted separately to the destination, frequently over different paths at different times. At the destination, the packets are electronically reassembled. This method may eventually be used for voice transmission.

PBX (Private Branch Exchange)—Customer premises equipment for switching intra-premises calls and connecting them to the telephone network. PBX is synonymous with PAX or Private Automatic Exchange.

PCM (Pulse Code Modulation)—A modern transmission system that sends voice signal information as a sequence of binary pulses representing numbers describing the signal amplitude on an instantaneous basis. The system is in use almost everywhere.

PIXEL—The smallest element of information on a video display.

PLAN OF REORGANIZATION—See POR.

POP (Point of Presence)—Locations established in each LATA where inter-LATA carriers connect with the local network.

POR (Plan of Reorganization)—A document prepared by AT&T to meet the requirements of the modification of final judgment, detailing exactly how divestiture would be accomplished. It was prepared in 1982 and approved with minor changes by Judge Greene.

PRIVATE BRANCH EXCHANGE—See PBX.

PROTOCOL—A procedure or set of directions providing for the orderly flow of information between information systems devices, such as data transmission terminals. This can also include a description of how data should be formatted, and the proper use of routing codes and addresses. It allows the devices to "talk" to each other.

RECEIVER—A unit of equipment designed to accept communications signals from a transmitter.

REGENERATION—A process used in digital transmission so that information pulses that have been attenuated and distorted in transmission can be recreated and made recognizable. While analogous to amplification in analog systems, it differs in that at each regeneration point, a new pulse is created and the old distortions are not carried forward.

REGISTER—A unit of equipment in common control switching systems—both electromechanical and electronic—receiving and recording information from a customer or an incoming trunk (such as the number being dialed) and then forwarding it to other equipment used to complete the call. It serves as a buffer between the relatively slow speed of a human and the high speed of electronic processing.

SAMPLING—A process used in digital transmission of analog signals that periodically measures, or samples, the signal. This sample is then converted into digital form for transmission.

SCANNER—A unit in an electronic switching system that

examines on a sequential basis the condition of each line in the office to see if there has been a change since the last examination. A change indicates something is happening concerning a call, such as the line has gone off-hook, on-hook, or there is dialing.

SIGNAL—The electrical representation of communications intelligence. It may represent information in voice, data, or video form.

SIGNALING—The transmission of information other than voice signals needed to set up a call. These include dial pulses and on- and off-hook signals. Traditionally signaling has been carried on the same circuit that carries the voice, but in newer systems it is transmitted separately for greater efficiency. See CCIS.

SINGING—An undesirable condition encountered on long voice frequency circuits with impedance irregularities. Under some conditions, a positive feedback path exists for certain frequencies, causing the generation of a high intensity tone that makes communication difficult or impossible. This is seldom a problem in modern circuit design.

SINGLE POINT OF CONTACT—See SPOC.

SOFTWARE—The detailed, step-by-step instructions used to operate electronic switching systems or computers. Physically, software programs are most often stored in solid state memory or on magnetic disks and provide the logical information for directing switching systems.

SOLAR CELL—A device that converts light into electrical energy.

SOLID STATE ELECTRONICS—A body of knowledge encompassing the use of semiconductors—more often than not, silicon—as circuit elements of all types. The transistor is the key element in this science. Virtually all progress in electronics in the last 30 years has been in the solid state field.

SPACE DIVISION—A type of switching system that provides a separate physical path through the machine for each simultaneous call. This is contrasted to a time division system where a common path is shared by many conversations, but separated into individual time slots.

SPOC (Single Point of Contact)—An organizational arrangement mandated by the modification of final judgment to coordinate national security requirements. It has been incorporated into Bellcore.

STATION EQUIPMENT—Equipment on a customer's premises. Used interchangeably with terminal equipment and customer premises equipment.

STEP BY STEP—An older type of electromechanical switching system that generally lacks central control and memory and operates on individual dial pulses as they are received. Now obsolete and used only in the smallest central offices, where it is being rapidly phased out.

STEREO SOUND—A system for high quality sound reproduction using separated microphone placements and speakers.

STORE AND FORWARD—See MESSAGE SWITCHING.

STORED PROGRAM—A set of logical instructions—software—stored in solid state memory or on magnetic disks, directing the operation of a central office.

SUPERVISION—Signals required to monitor the status of a call, such as on-hook, off-hook, etc. See SIGNALING.

SWITCHING—The interconnection of customer lines, trunks, or lines with trunks. The function is performed in central offices on machines called switching systems or switchers.

T1 CHANNEL—A basic building block in digital transmission. A T1 channel operates at a rate of 1.54 million bits per second, and can be used to transmit 24 voice grade telephone channels, or various combinations of other information adding up to the channel's bit rate capacity. T1 channels can be "stacked"—combined—for more efficient transmission at higher bit rates over broadband media such as optical fiber or microwave radio.

TERMINAL EQUIPMENT—Equipment at the ends of a circuit, usually on customer's premises. Also referred to as CPE and station equipment.

TIME DIVISION—See SPACE DIVISION.

TIME SLOT—A very short interval of time—measured in microseconds—assigned to a particular communications channel in a time division switching machine.

TOLL CENTER—The next level in the switching hierarchy above the end office. Customers' lines don't usually connect directly to these centers. Toll centers are also referred to as Class 4 offices.

TRAFFIC—The flow or volume of telephone calls. The discipline of traffic engineering includes determining the quantities of equipment and trunks needed for a specified average grade of service.

TRANSISTOR—A semiconductor device—usually made of silicon—which has replaced the vacuum tube in virtually all applications. Advantages over the vacuum tube include low cost, low power requirements, small size, and long life. Transistor functions are commonly provided as part of the multiple circuit elements on silicon chips. See CHIP and LSI.

TRANSMISSION—The art of carrying telecommunications signals from one point to another over wire, cable, radio, and optical fiber.

TRANSMITTER—A circuit unit designed to send communications signals to a receiver.

TRUNK—A circuit carrying telephone traffic that interconnects central offices. Any telephone call between customers served by different central offices requires at least one trunk. Often a number of trunks are connected in tandem to build a connection through a hierarchy of switchers.

VACUUM TUBE AMPLIFICATION—The process of strengthening attenuated signal levels using vacuum tube circuitry. This has been almost completely replaced by transistor technology.

VALUE OF SERVICE—A concept holding that prices should be based on what a service is worth to the customer rather than on the costs of providing it. This concept long governed pricing in telecommunications services. Using it, rural service, for example, was priced similarly to urban service, despite it often costing more to provide. The value of service concept is not viable in a competitive environment.

VERTICAL INTEGRATION—The combination, under one management, of functions from research through manufacture, installation and service. The Bell System worked on this basis for years with Bell Telephone Laboratories, Western Electric, and the Bell Operating Companies under AT&T management.

VSLI (Very Large Scale Integration)—LSI on a larger scale. See LSI.

VOICE CHANNEL—A communications path capable of carrying one voice quality circuit, typically with a frequency range up to about 3,500 hertz.

VOICE FREQUENCY FACILITIES—Copper pairs used to carry a single voice channel without modulation. Prior to 1940, almost 100 percent of telephone circuits fell into this category. Now voice frequency facilities are used only for short distances because using pulse code modulation—PCM—to combine many circuits saves copper and is therefore more economical. See PCM.

VOICEMAIL—A system for storing and forwarding voice messages to predetermined destinations at preset times.

WAVEGUIDE—A structure that transmits high frequency electrical signals in the form of waves. Copper pipes are used in this way to carry microwave electromagnetic signals, usually between transmitters or receivers and antennas. Optical fibers are similarly used to carry light waves over long distances.

BIBLIOGRAPHY

BOOKS

Schlesinger, Leonard A.; Dyer, David; Clough, Thomas N.; and Landau, Diane. *Chronicles of Corporate Change*. Lexington, Massachusetts and Toronto: Lexington Books, 1987.

Temin, Peter; and Galambos, Louis. *The Fall of the Bell System*. Cambridge: Cambridge University Press, 1987.

Kleinfield, Sonny. *The Biggest Company on Earth: A Profile of AT&T*. New York: Holt, Rinehart, and Winston, 1981.

Brooks, John. *Telephone: The First Hundred Years*. New York: Harper & Row, 1975.

Mabon, Prescott C. *Mission Communications: The Story of Bell Laboratories*. Murray Hill, New Jersey: Bell Laboratories Incorporated, 1975.

Boettinger, H. M. *The Telephone Book: Bell, Watson, Vail and American Life 1876–1976*. Croton-on-Hudson, New York: Riverwood Publishers Limited, 1977.

Tunstall, W. Brooke. *Disconnecting Parties: Managing the Bell System Break-up*. New York: McGraw-Hill Book Company, 1985.

von Auw, Alvin. *Heritage & Destiny: Reflections on the Bell System in Transition*. New York: Praeger Publishers, 1983.

Martin, James. *Future Developments in Telecommunications*. Englewood Cliffs, New Jersey: Prentice-Hall Inc., 1971.

Pierce, John R. *Signals: The Telephone and Beyond*. San Francisco: W. H. Freeman and Company, 1981.

Coll, Steve. *The Deal of the Century: The Breakup of AT&T*. New York: Atheneum, 1986.

Phillips, Almarin. *The Impossibility of Competition in Telecommunications: Public Policy Gone Awry*, from *Regulatory Reform in Public Utilities*. Lexington: Lexington Books, 1982.

Bell Laboratories Technical Staff. *A History of Engineering and Science in the Bell System* (seven volumes). Murray Hill, New Jersey: Bell Laboratories Incorporated, 1975–1984.

PAPERS, ARTICLES

Bolling, Colonel George H. *AT&T—Aftermath of Antitrust: Preserving Positive Command and Control.* Washington: National Defense University Press, 1983.

Reinman, Colonel Robert A. *National Telecommunications Emergency Policy: Who's In Charge?.* Washington: National Defense University Press, 1984.

Dittberner, Donald L. *Interconnection Action Recommendations.* Washington: U.S. Department of Commerce, 1970.

Bell, Trudy E. *Hello Again: The Future of Telecommunications.* Institute of Electrical and Electronic Engineers Spectrum Magazine 22, November 1985:44–113.

Kraus, Constantine Raymond. *One Communications System for Mankind.* Institute of Electrical and Electronic Engineers Spectrum Magazine 22, June 1985:8.

Sirbu, Marvin A. Jr. *A Review of Common Carrier Deregulation in the United States: Lessons for Foreign PTTs and Operating Authorities.* Cambridge, Massachusetts: Massachusetts Institute of Technology, 1982.

Duerig, Alfred W. *The Demise of the Telephone Network.* Public Utilities Fortnightly Magazine 117, January 1986:30–38.

Kahn, Alfred E. *The Uneasy Marriage of Regulation and Competition.* Telematics: September 1984.

Kraus, Constantine Raymond. *Regulation of Communications is not Coping With Technology.* Telephone Engineer and Management Magazine: March 1973.

Colter, Cyrus J. *The Consequences of Competition.* Bell Telephone Magazine: October 1984.

Kraus, Constantine Raymond. *Social Consciousness in Communications Engineering.* Institute of Electrical and Electronic Engineers Communications Society: May 1976.

Stavro, Barry. *Therefore Be Bold.* Forbes Magazine: September 1985.

Gold, Howard. *The Great Telephone Election.* Forbes Magazine 136: November 1985.

Morgan, M. Granger; and Sirbu, Marvin. *Divestiture and Deregulation: an Untidy Process.* Institute of Electrical and Electronic Engineers Spectrum Magazine 22, December 1985:8.

Kraus, Constantine Raymond. *Proposal for a New Nationwide Communications Public Utility Service.* Telephone Engineer and Management Magazine: August 1972.

The Public Utility Commission of Prince Edward Island, Canada—
Decision on Interconnect—1980.

Southern Pacific Communications Company vs. AT&T. United
States District Court, District of Columbia, Judge Charles R.
Richey. Federal Supplement 566.

PATENTS

C. R. Kraus, U.S. Patent No. 3,728,486—"Voicegram Service for a Telephone System." (Patents also in France, United Kingdom, Canada
and Japan.)

C.R. Kraus, U.S. Patent No. 3,920,908—"A Buyer Credit System."

C.R. Kraus, U.S. Patent No. 4,660,220—"A No Answer Tone for the
Telephone System."

Index

ABOUT THE AUTHORS

Constantine Raymond Kraus has spent much of his working life as an engineer and adminstrator with the Bell System. He has been intimately connected with every facet of communications. When he retired from Bell, he founded and headed Consulting Communications Engineers, Inc., of Villanova, bringing together a brain trust of 200 retired scientists and engineers representing 6,000 years of experience and 500 patents. The 20-year-old company has worked for both government and industry. Mr. Kraus also edits the CCE newsletter, which has a wide circulation in the industry.

Alfred W. Duerig was employed by the Bell Telephone Company of Pennsylvania for thirty-seven years; during the last seventeen he headed the engineering department. Before that he served in the United States Navy in various communications positions. He is a registered professional engineer and also a senior member of the Institute of Electrical and Electronics Engineers.